Imaging Mass Spectrometry

Mitsutoshi Setou
Editor

Imaging Mass Spectrometry

Protocols for Mass Microscopy

 Springer

Editor
Mitsutoshi Setou M.D., Ph.D.
Professor and Chairman
Molecular Imaging Frontier Research Center
Department of Molecular Anatomy
Hamamatsu University School of Medicine
1-20-1 Handayama, Higashi-ku, Hamamatsu
Shizuoka 431-3192, Japan
setou@hama-med.ac.jp

This book is based on the Japanese original, *Shitsuryo Kenbikyo Ho: Imaging Mass Spectrometry Jikken Protocol*, M. Setou, editor; published by Springer Japan, 2008.

ISBN 978-4-431-09424-1   ISBN 978-4-431-09425-8 (eBook)
DOI 10.1007/978-4-431-09425-8
Springer Tokyo Berlin Heidelberg New York

Library of Congress Control Number: 2009939108

Springer is part of Springer Science+Business Media (www.springer.com)

Preface

Imaging mass spectrometry (IMS, also known as mass spectrometry imaging) is a technique to visualize the spatial distribution of, e.g., compounds, biomarkers, metabolites, peptides, or proteins by their molecular masses. Recently, research groups around the world have developed IMS, and pharmaceutical companies have started using IMS technology for purposes such as drug ADME (absorption, distribution, metabolism, excretion) analyses to study pharmacokinetics and the medical properties of drugs. This use of IMS technology will continue to expand in the areas of drug research and disease-related biomarker screening. Consequently, the number of researchers interested in the practical use of IMS is increasing both in academic and in industrial fields.

Under these circumstances, there is a real need for practical knowledge and techniques for the application of IMS. This book is intended for beginners as well as for researchers who have relatively little experience in the field. Detailed practical protocols are covered, so readers will immediately be able to perform IMS. We hope this book lays the groundwork for future progress in IMS technique and ultimately makes a contribution to progress in human health.

Mitsutoshi Setou

Contents

Contributors

Soeren-Oliver Deininger (*Chapters 7 and 15*)
Bruker Daltonik GmbH, Bremen, Germany

Masaru Furuta (*Chapter 16*)
Applications Development Center, Analytical Applications Department,
Analytical and Measuring Instruments Division, Shimadzu Corporation,
Kyoto, Japan

Naoko Goto-Inoue (*Chapter 13*)
Department of Molecular Anatomy, Hamamatsu University School of Medicine,
Hamamatsu, Shizuoka, Japan

John S. Hammond (*Chapter 18*)
Physical Electronics, Chanhasse, MN, USA

Takahiro Hayasaka (*Chapter 8*)
Department of Molecular Anatomy, Hamamatsu University School of Medicine,
Hamamatsu, Shizuoka, Japan

Daisuke Horigome (*Chapters 5 and 6*)
Mitsubishi Kagaku Institute of Life Sciences, Machida, Tokyo, Japan

Naofumi Hosokawa (*Chapter 9*)
Department of Bioscience and Biotechnology, Tokyo Institute of Technology,
Yokohama, Kanagawa, Japan
and
Mitsubishi Kagaku Institute of Life Sciences, Machida, Tokyo, Japan

Koji Ikegami (*Chapter 12*)
Department of Molecular Anatomy, Hamamatsu University School of Medicine,
Hamamatsu, Shizuoka, Japan

Tetsuo Kokaji (*Chapter 14*)
Applied Biosystems Japan, Chuo-ku, Tokyo, Japan

Motoji Oshikata (*Chapter 17*)
Waters Corporation, Shinagawa-ku, Tokyo, Japan

Martin Schuerenberg (*Chapter 7*)
Bruker Daltonik GmbH, Bremen, Germany

Mitsutoshi Setou (*Chapters 1–6, 9, 10–13, 16 and 17*)
Mitsubishi Kagaku Institute of Life Sciences, Machida, Tokyo, Japan
and
Department of Molecular Anatomy, Hamamatsu University School of Medicine,
Hamamatsu, Shizuoka, Japan

Yuki Sugiura (*Chapters 2, 4–6, 9, 10, 12, 16 and 17*)
Department of Bioscience and Biotechnology, Tokyo Institute of Technology,
Yokohama, Kanagawa, Japan
and
Mitsubishi Kagaku Institute of Life Sciences, Machida, Tokyo, Japan
and
Department of Molecular Anatomy, Hamamatsu University School of Medicine,
Hamamatsu, Shizuoka, Japan

Takao Taki (*Chapter 13*)
Institute of Biomedical Innovation, Otsuka Pharmaceutical Co. Ltd.,
Kawauchi-cho, Tokushima, Japan

Hyun Jeong Yang (*Chapter 12*)
Department of Bioscience and Biotechnology, Tokyo Institute of Technology,
Yokohama, Kanagawa, Japan
and
Department of Molecular Anatomy, Hamamatsu University School of Medicine,
Hamamatsu, Shizuoka, Japan

Ikuko Yao (*Chapter 3*)
Department of Medical Chemistry, Kansai Medical University, Moriguchi,
Osaka, Japan

Naohiko Yokota (*Chapter 17*)
Waters Corporation, Shinagawa-ku, Tokyo, Japan

Nobuhiro Zaima (*Chapter 11*)
Department of Molecular Anatomy, Hamamatsu University School of Medicine,
Hamamatsu, Shizuoka, Japan

Part I
Innovation of Imaging Mass Spectrometry

Chapter 1
IMS as an Historical Innovation

Mitsutoshi Setou

Abstract The development of mass spectrometry has recently entered a new phase. The previous limitation of mass analyses of biomolecules confined the imagination of researchers to a single location from which samples were gathered; however, an innovation in mass spectrometry has now enabled the provision of additional two-dimensional (2D) axes of recognition on tissue sections. The innovatory technology – imaging mass spectrometry – can visualize the distribution of molecules. It has attracted a great deal of attention in the analyses of drug effects, in screening of drugs, and in support for medical diagnoses. In this chapter, we describe imaging mass spectrometry as an historical innovation and the significance of publishing this protocol book. I strongly believe that this book provides a critical basis for future studies on imaging mass spectrometry.

Imaging mass spectrometry (IMS) is an imaging technique based on mass spectrometry. Images are reconstructed from the mass spectrum charts from thousands of spots from biological tissues to show the distribution of various molecules. The ionization techniques often used are MALDI (matrix-assisted laser desorption ionization), DESI (desorption electrospray ionization), or SIMS (secondary ion mass spectrometry) [1].

SIMS imaging, which uses the second electronic ion, was theoretically invented in 1949 by Herzog and Viehb of Vienna University. The first SIMS device was completed by Liebel and Herzog in 1961, and it was utilized for the surface analysis of metals, which is its widest use at present. The high resolution obtained in this imaging method was comparable to that required for microscopic levels; however, it is not suitable for analyses of biological macromolecules

M. Setou (✉)
Mitsubishi Kagaku Institute of Life Sciences, 11 Minamiooya, Machida, Tokyo, Japan and
Department of Molecular Anatomy, Hamamatsu University School of Medicine,
1-20-1 Handayama, Higashi-ku, Hamamatsu, Shizucka 431-3192, Japan
e-mail: setou@hama-med.ac.jp

M. Setou (ed.), *Imaging Mass Spectrometry: Protocols for Mass Microscopy*,
DOI 10.1007/978-4-431-09425-8_1, © Springer 2010

because the second electronic ion beam breaks the structure. In 1997, the group of Dr. Caprioli of Vanderbilt University presented the first paper regarding MALDI imaging [2]. DESI imaging is a relatively new technique, first reported by the group of Dr. Cooks [3]. Dr. Heeren originally reported the stigmatic type of IMS instrument [4]. Recently, our group first developed high-resolution IMS (mass microscopy), such as nanoparticle-assisted laser desorption/ionization (nano-PALDI) [5], for microscopic observation.

The major applications of IMS technology are in a pharmaceutical company, such as for pharmacokinetic monitoring, pharmacotoxicology, and pharmacometabolomes (Fig. 1.1). Although the image resolution of in vivo ADME (absorption, distribution, metabolism, and excretion) by current IMS technology is relatively lower than that obtained using isotope-labeled compounds, preparing for IMS-based experimentation is simpler and easier than for other methods, particularly when thousands of lead compounds need to be tested. IMS, with its higher image resolution, will contribute greatly to understanding of drug action mechanisms, and so they are expected to improve in the near future. Further, IMS is a strong and perhaps unique tool for performing so-called metabolome-mapping while using only a single-sample analysis, thus monitoring the metabolic behavior of thousands of molecules in terms of their quantitative and positional properties.

With the advantage of no labeling, IMS has opened a new frontier in medical and clinical applications. Other molecular imaging techniques such as green fluorescent protein (GFP) labeling or immunohistochemistry require labeling. Lipids and low molecular weight compounds in tissue sections cannot be observed with those conventional microscopic and electron microscopic techniques; therefore, no distribution map of these molecules in tissue structure has to date been described in the scientific literature or medical textbooks. IMS is making possible a characteristic distribution map of lipids (Fig. 1.2), thus also making a major impact on lipid research.

Ordinary techniques for proteomic and metabolomic analysis cannot be applied to biopsies because most of the samples are extremely small to minimize patient burden. For example, if you have a 1-mm^3 sample of biopsy tissue, 3 µl buffer solution

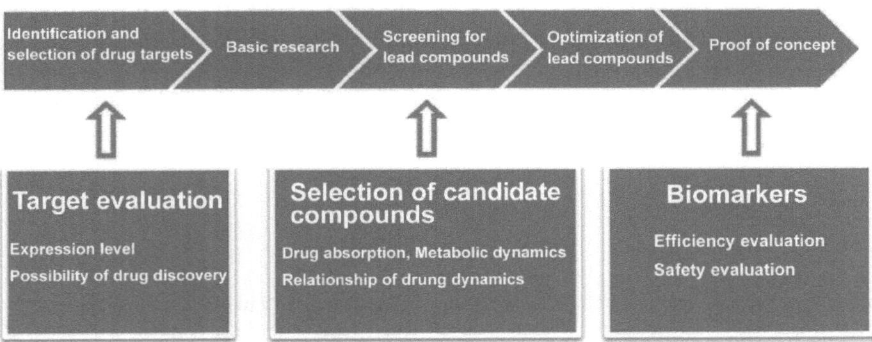

Fig. 1.1 Processes of new drug development and areas of application of mass microscopy

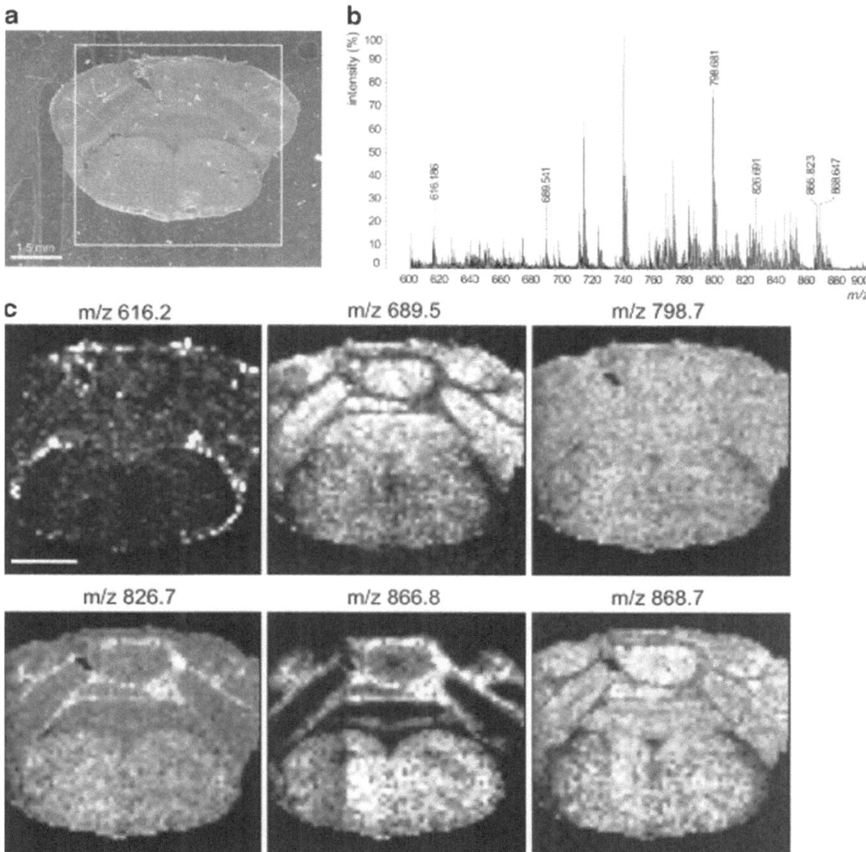

Fig. 1.2 Characteristic lipid distribution in a rodent brain (Reprinted from Shimma et al., Anal Chem 80(3):878–885.)

is used to dissolve it. Such minute amounts of sample solution would be lost by nonspecific adsorption during subsequent preparation by centrifugation and chromatography. In contrast, when using the IMS technique, the sample would first be flash-frozen, and several μm-thick tissue sections would then be obtained to perform an accurate metabolomic analysis [6, 7]. IMS is expected to become a standard method for reviewing clinical metabolomes because its technique is simple and easy. The mass microscope is now standing alongside computed tomography (CT), positron emission tomography (PET), and magnetic resonance imaging (MRI) in our hospital at Hamamatsu University School of Medicine (Table 1.1).

Any innovation is based on previous efforts. Isaac Newton refers to these achievements as "standing on the shoulders of giants." Within various great innovations, several developments sometimes appear to embody a "great leap" from older discoveries and newborn innovations. Figure 1.3 is my personal view of innovations

Table 1.1 Comparison of characteristics of molecular visualization methods

	MRI	NIRF	PET	Biolu.	WBAL	Opt.	IMS
In vivo	✓	✓	✓	✓	✗	✗	✗
Sensitivity	μM	nM	pM	nM	nM	nM	μM
Resolution	50 μm	5 mm	5 mm	1 mm	10 μm	1 μm	50 μm
Time required	min	min	min	s	day	min	min
Labeling	✓	✓	✓	✓	✓	✓	✓
Dimension	3D	2(3)D	3D	2(3)D	2D	2D	2D
Cost	¥¥¥	¥	¥¥¥¥	¥	¥¥	¥	¥¥

MRI magnetic resonance imaging, *NIRF* near-infrared fluorescence imaging, *PET* positron emission tomography, *Biolu.* bioluminescence, *WBAL* whole-body autoradioluminography, *Opt.* optical, *IMS* imaging mass spectrometry

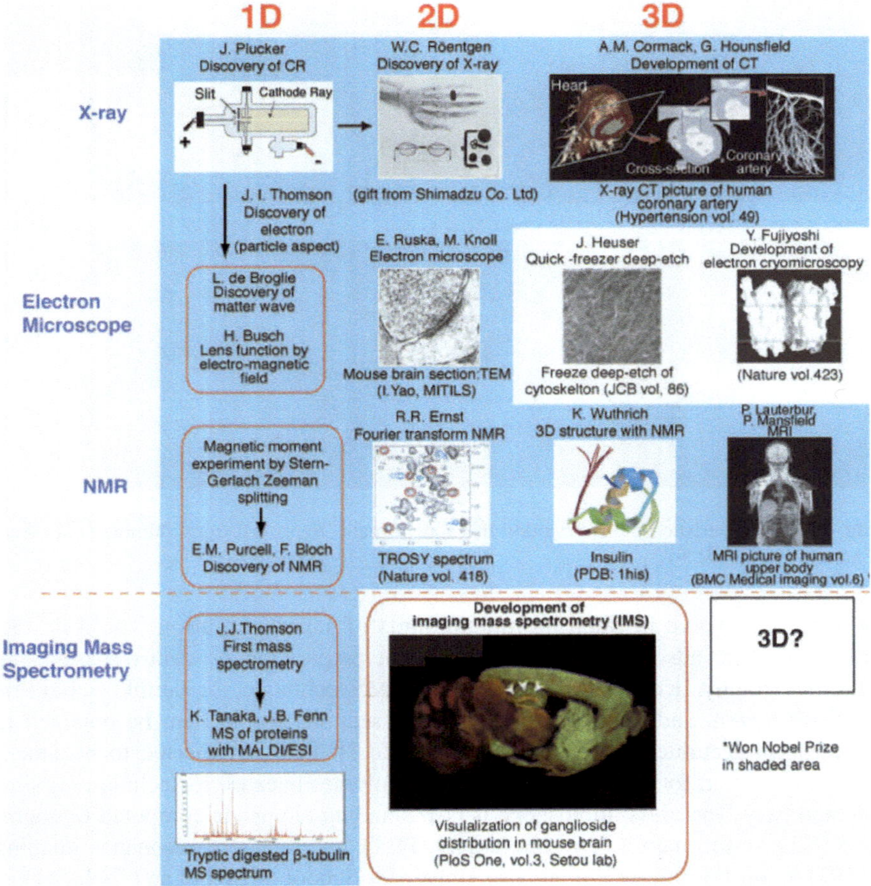

Fig. 1.3 Historical view of the development of analytical instruments. The progress of imaging technology has been categorized by its technology (*longitudinal axis*) and its analytical dimensions (*horizontal axis*). *NMR* nuclear magnetic resonance. Work shown with *blue area* has received a Nobel Prize. *CR* cathode ray, *NMR* nuclear magnetic resonance, *CT* computed tomography, *TEM* transmission electron microscopy

made in modern biotechnology. The "ground-breaking steps" taken en route to generating innovation – "quantum leaps" – emerge and demonstrate certain steps in the dimensions of analysis. Cathode-ray equipment, for example, evolved into X-ray photographs and, in turn, into CT scans. The electron beam has been used in lieu of an electron microscope. Zeeman splitting has been applied to two-dimensional (2D) nuclear magnetic resonance (NMR) images and then to magnetic resonance (MR) images. The steps taken in the process of innovation match the concomitant increases to analytical dimensionality. The first dimension of analysis is based on basic physics. In the second dimension, an innovation is further developed by means of contributions made by the field of chemistry; subsequently, it is applied to biology and/or medical procedures. Obviously, the next breakthrough for the IMS is to make it 3D (three-dimensional). Toward this direction, we recently have developed atmosphere pressure mass microscopy.

References

1. Shimma S, Setou M (2005) Review of imaging mass spectrometry. J Mass Spectrom Soc Jpn 53:230–238
2. Caprioli RM, Farmer TB, Gile J (1997) Molecular imaging of biological samples: localization of peptides and proteins using MALDI-TOF MS. Anal Chem 69(23):4751–4760
3. Takáts Z, Wiseman JM, Gologan B, Cooks RG (2004) Mass spectrometry sampling under ambient conditions with desorption electrospray ionization. Science 306:471–473
4. Maarten Altelaar AF, Taban IM, McDonnell LA, et al. (2007) High-resolution MALDI imaging mass spectrometry allows localization of peptide distributions at cellular length scales in pituitary tissue sections. Int J Mass Spectrom 260:203–211
5. Taira S, Sugiura Y, Moritake S, et al. (2008) Nanoparticle-assisted laser desorption/ionization based mass imaging with cellular resolution. Anal Chem 80(12):4761–4766
6. Shimma S, Setou M (2007) Mass microscopy revealed the distinct localization of heme B(m/z 616) in colon cancer liver metastasis. J Mass Spectrom Soc Jpn 55:145–148
7. Shimma S, Sugiura Y, Hayasaka T, et al. (2007) MALDI-based imaging mass spectrometry revealed abnormal distribution of phospholipids in colon cancer liver metastasis. J Chromatogr B Anal Technol Biomed Life Sci 855:98–103

Part II
Planning the Sample Preparation

Chapter 2
Guide to Planning the Sample Preparation Step

Yuki Sugiura and Mitsutoshi Setou

Abstract Matrix-assisted laser desorption/ionization (MALDI)-imaging mass spectrometry (IMS; also referred to as mass spectrometry imaging, MSI) enables the visualization of the distribution of variable biomolecules that have large and varied structures in tissue sections. However, because of such general versatility, the optimization of sample preparation procedure according to each analyte with distinct chemical and physical properties is an important task for IMS, similar to that in traditional MALDI-MS analyses. In this chapter, we briefly introduce the sample preparation strategies of current IMS. We describe the two major strategies for imaging of proteins/peptides and small molecules, such as lipids and drugs. We shall also introduce the representative application example.

2.1 Introduction

Traditional mass spectrometry (MS), particularly MALDI-MS, can be applied to the analysis of a wide range of molecules that have various physical and chemical features. In biology- and medicine-related fields, therefore, MALDI imaging MS have been applied to a visualization of the distribution of many kinds of biomolecules in cells and tissues. Because of such general versatility, the optimization of experimental protocols for sample preparation, the conditions of MS and choosing a data analysis method are important issues to consider. Particularly, the procedure for sample preparation is critical.

Y. Sugiura
Department of Bioscience and Biotechnology, Tokyo Institute of Technology,
4259 Nagatsuta-cho, Midori-ku, Yokohama, Kanagawa 226-8501, Japan

M. Setou (✉)
Department of Molecular Anatomy, Hamamatsu University School of Medicine,
1-20-1 Handayama, Higashi-ku, Hamamatsu, Shizuoka 431-3192, Japan
e-mail: setou@hama-med.ac.jp

M. Setou (ed.), *Imaging Mass Spectrometry: Protocols for Mass Microscopy*,
DOI 10.1007/978-4-431-09425-8_2, © Springer 2010

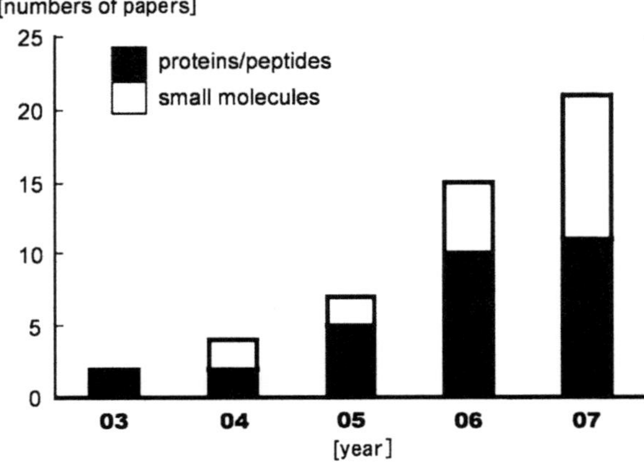

Fig. 2.1 Results of the PubMed search using "Imaging Mass Spectrometry" as key words

In the following sections, we describe two major sample preparation strategies, namely, IMS for proteins/peptides and for small organic molecules. To date, most of the reports concerning MALDI-IMS are with regard to the detection and imaging of proteins or peptides. On the other hand, the amount of research regarding the detection and imaging of small organic molecules has been growing recently. Figure 2.1 shows the result of a PubMed search using "imaging mass spectrometry" as the key word (excepting reviews). Reports were subdivided into groups according to the analytes. Notably, the number of reports regarding the IMS of small compounds gradually increased to encompass half the published works in 2007. We believe that these two major fields will continue to grow.

Below we describe detailed sample preparation strategies in each field and introduce a representative example of research applications. For further information, the inquisitive reader is directed to the excellent references cited herein. We hope that this chapter will assist researchers as they choose the appropriate method for pretreating samples for IMS.

2.2 Characteristics of MALDI in Direct Tissue Mass Spectrometry

In MALDI-IMS, in which the tissue sample is directly measured, researchers should pay special attention to the fact that samples are extremely complex mixtures of biomolecules. Because tissues and cells are subjected to MALDI-IMS, the sample cleanup procedure is limited, whereas in traditional MS research, analyte molecules are generally extracted and separated from such crude samples by using gas chromatography or high performance liquid chromatography. Generally when those crude samples are subjected to MS, numerous molecular species compete for ionization; eventually, molecules that are easily ionized preferentially reach the detector while suppressing the ionization of other molecules. This phenomenon is called the "ion suppression effect" [1–3]. Figure 2.2 shows IMS results of a peptide solution (0.5 µl 100 nM ACTH) spotted on both an Indium thin oxide (ITO)-glass slide surface and a brain section and clearly demonstrates that the ion suppression effect severely occurs on the tissue surface; the spotted peptide was detected only from the ITO-surface.

As a practical example, when a mouse brain section to which 2,5-dihydroxybenzoic acid (DHB) has been applied as a matrix is subjected directly to MS in positive ion detection condition, strong peaks are observed in the mass region of $700 < m/z < 900$; they are mainly derived from phospholipids [4]. Signals derived from proteins, meanwhile, are scarcely detected in $m/z > 3,000$ [5], because phospholipids ionize much more efficiently than proteins and suppress protein/peptide ionization.

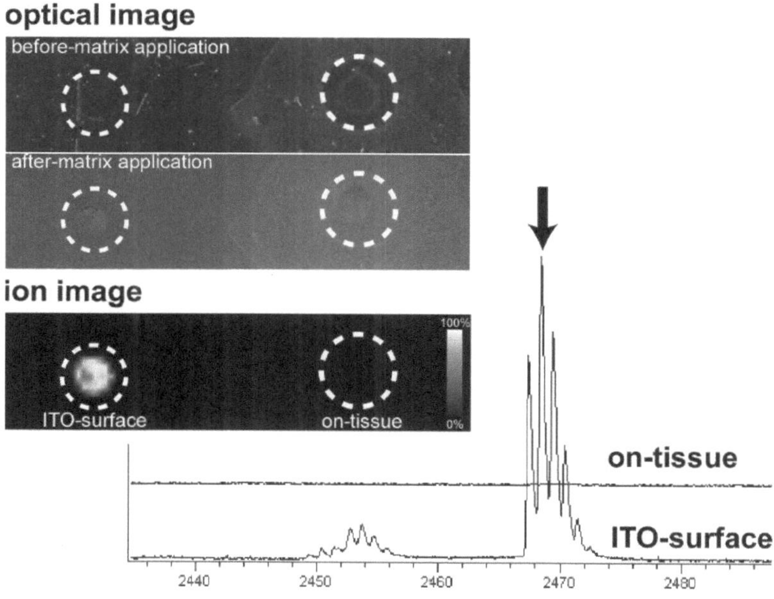

Fig. 2.2 Example of ion suppression effect on the tissue surface. A peptide solution (0.5 µl 100 nM ACTH) was spotted on both an ITO-glass slide surface and a brain section. After spraying the matrix solution [10 mg ml⁻¹ alpha-cyano-4-hydroxy cinnamic acid (α-CHCA), 50% acetonitrile, 0.1% trifluoroacetic acid (TFA)], the spotted peptide was visualized by IMS

So, one may ask: what kinds of molecules are dominantly ionized via the MALDI imaging? In general, we adopt the following criteria:

1. Large amount
2. Contains a chemical structure that can be easily charged (e.g., basic groups such as amine- and nitrogen-containing heterocyclic compounds)
3. Has a chemical structure that absorbs UV light (e.g., cyclic organic compounds) (when a UV laser is used for ionization)
4. Easily vaporized
5. Co-crystallizes easily with the matrix

When an infrared (IR) laser is used, the compounds that absorb IR are easily ionized [6].

In brain tissues, phospholipids satisfy most of these criteria; first, a large amount – more than 60% in dry weight – of a mouse brain consists of lipids. Second, phospholipids have easily ionized structures; polar lipids, particularly phosphatidylcholine, contain a phosphate and trimetylamine group both of which facilitate addition of charge to the molecule [7]. Therefore, for detecting/imaging proteins and peptides, one needs to undertake sample preparation steps in which such lipids are removed. To this end, tissue sections should be rinsed with organic solvent [5, 8, 9]. This sample preparation step can lead to ionization of proteins and peptides, which are the second most abundant substance in cells, when the tissues are directly analyzed by MS and signals in the range of $3,000 < m/z < 20,000$ can clearly appear. As this example demonstrates, the sample preparation step has a critical role in the successful IMS.

2.3   IMS for Proteins and Peptides

As already described, rinsing tissue sections is a key component of the sample preparation step for the analysis of proteins and peptides (Fig. 2.3).

Rinsing the sections enhances the sensitivity required for protein/peptide detection by eliminating small molecules, particularly phospholipids, from the tissues [5]. Regarding this sample preparation step, Lemaire et al. have examined the use of variable organic solvents in the rinsing step to enhance the efficiency of protein detection in tissue [5]. Figure 2.4 show mass spectra of $500 < m/z < 960$, obtained from mouse brain sections without rinsing or rinsed with chloroform, acetone, or hexane (Fig. 2.4a, 4b, 4c, or 4d, respectively). Although signals in the vicinity of m/z 700 predominately appear in untreated tissues, they are abolished after rinsing. Conversely, the number of peaks derived from proteins at a high molecular weight (i.e., $5,000 < m/z$) increased more than 40% after rinsing (Table 2.1). This treatment also helps remove salts, a detrimental factor that otherwise interferes with the matrix–analyte co-crystallization process and thus degrades sensitivity, while also complicating the spectrum assignment by producing both protonated and cationated molecular ions [8].

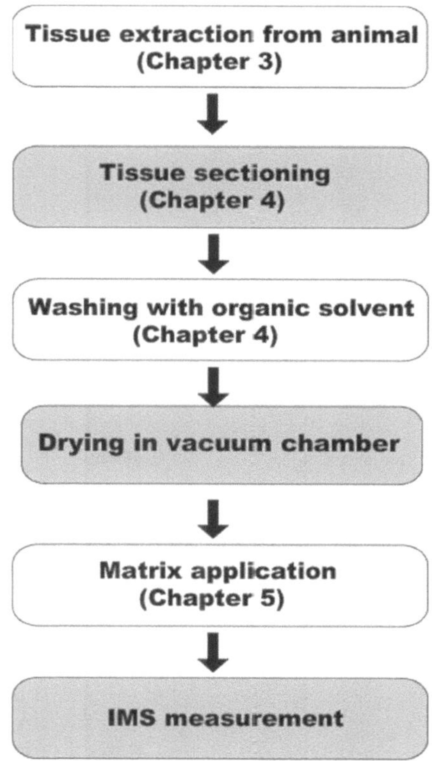

Fig. 2.3 A scheme for protein and peptide measurement

Choosing the matrix is another important step. General recommendations in traditional MALDI-MS have been well studied [10]. In MALDI-IMS, regarding this issue for large protein measurement, sinapic acid (SA) is frequently used as a suitable matrix, rather than α-cyano-4-hydroxycinnamic acid (CHCA) or DHB (Fig. 2.5) [8]. Schwartz et al. [8] have reported that the use of SA enables the detection of more peaks derived from proteins with high molecular weights, with very little background noise, than do other matrices. On the other hand, CHCA can be used for the detection of relatively small peptides (i.e., $<m/z$ 5,000) [11]. Nevertheless, performing a preliminary test for the selection of matrix – using a reference compound – is recommended for successful analyte detection in IMS. Further details regarding the selection and coating of the matrix are described in Chap. 5.

The efficient extraction of proteins/peptides from tissues to the matrix solution is also important factor in successful IMS analysis. In tissue MS, only analytes extracted from tissue and are well co-crystallized with matrices can be ionized. Thus, by enhancing protein/peptide extraction, the addition of detergents to the matrix solution is expected to increase analyte ion signals because detergents promote to extract protein/peptide from tissue and to form co-crystallization [12].

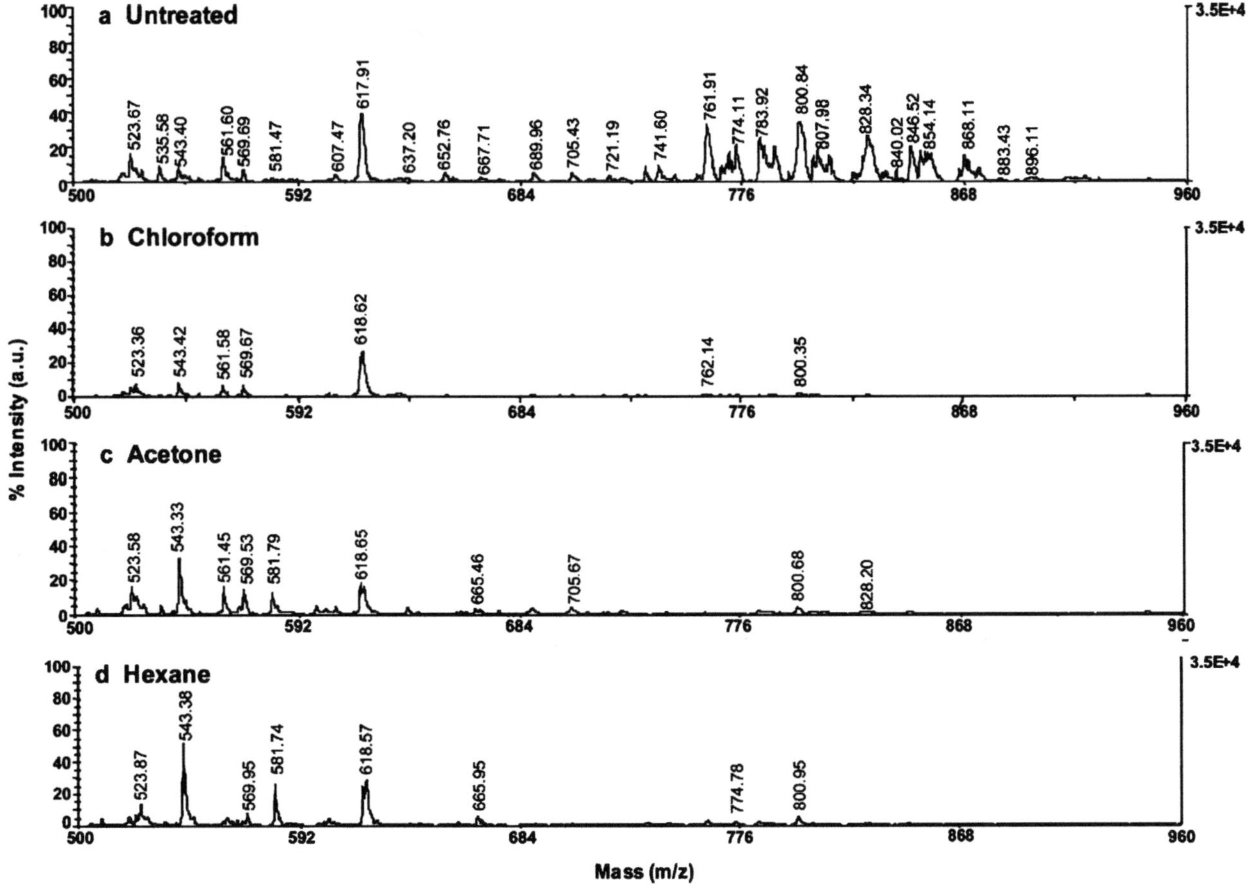

Fig. 2.4 Mass spectra (500<m/z<960) obtained from mouse brain sections without pretreatment (**a**) or pretreated with chloroform (**b**), acetone (**c**), or xylene (**d**). α-Cyano-4-hydroxycinnamic acid (CHCA) was used as a matrix [5] (Reprinted from Lemaire et al., Anal Chem 78(20):7145–7153.)

Table 2.1 Improvement of signals derived from proteins and peptides after rinsing with organic solvent (Reprinted from Lemaire et al., Anal Chem 78(20):7145–7153.)

Treatment	n	No. of detected compounds	Standard deviation (%)	Increase in detection (%)
Chloroform	10	81	22	34
Hexane	5	75	28	25
Toluene	5	68	22	13
Xylene	5	86	13	44
Acetone	5	64	29	7
Untreated	10	60	34	0

n number of experiments

Average number of detected compounds, standard deviation, and calculated increased detection for peptides/proteins of $m/z>5,000$ determined from the mass spectra recorded on untreated rat brain sections vs. organic solvent-treated sections [5]

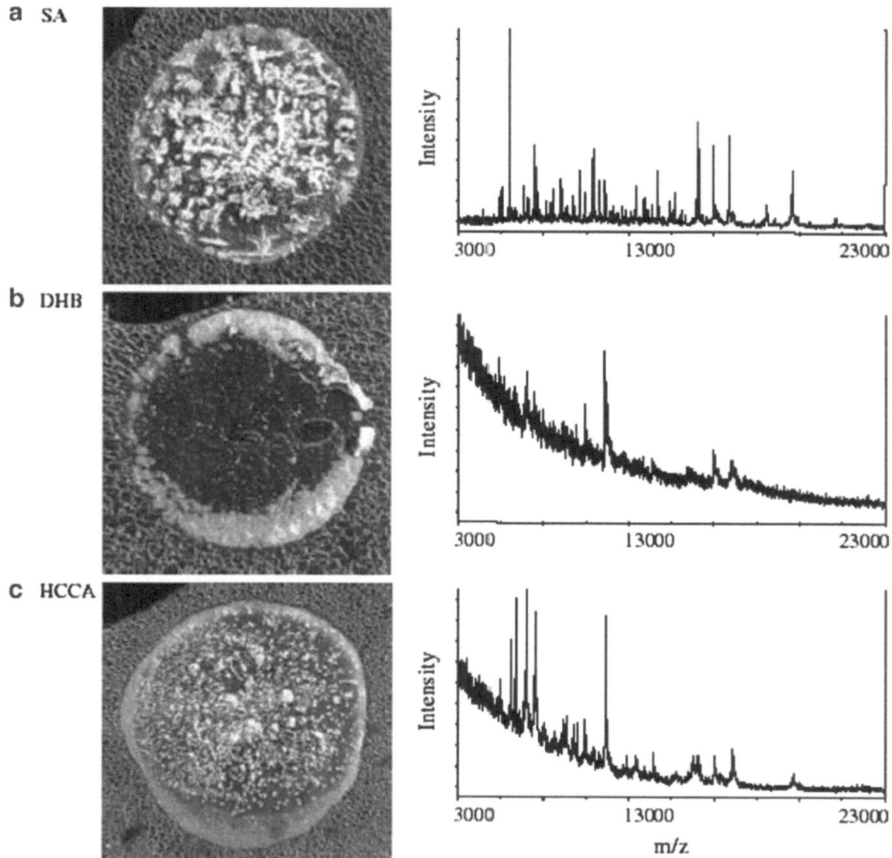

Fig. 2.5 Crystals of matrices and spectra in analyses in which SA (**a**), DHB (**b**), or HCAA (**c**) is used as a matrix are shown [8] (Reprinted from Schwartz et al., J Mass Spectrom 38:699–708.)

Although most detergent reagents interfere with the crystallization of matrices and degrade signals, *n*-octyl-glucoside, which does not decrease sensitivity, is used to enhance the extraction of proteins and peptides [12, 13]. The method of matrix application also influences analyte extraction. Compared to a spray-coating of matrix solution, the deposition of matrix solution droplets by a pipette or an automatic dispenser increases signal sensitivity [8] because the large droplets of the solution enhance protein extraction rather than those of the spray-coating. Chapter 5 provides details about methods of matrix application.

2.3.1 On-Tissue Digestion Method

As observed in sodium dodecyl sulfate-polyacrylamide gel electrophoresis (SDS-PAGE) of crude extracts from cells or tissues, there are many proteins much larger than 40 kDa, and these proteins have important biological functions. However, in the current IMS, the upper limit of protein detection is approximately 40 kDa. Another limitation is that some proteins such as membrane proteins are difficult to be extracted from tissues, and provide low sensitivity for the detection.

To solve these problems, the on-tissue digestion method, in which proteins are denatured and digested by enzymes, has been developed [14–16]. When large proteins are cleaved by this method, they can be detected by MS as digested proteins, mainly in the mass range of $900 < m/z < 3,000$. Protein digestion also enables the efficient identification of molecular species by tandem MS (Fig. 2.6) [14, 16]. Furthermore, this process was enhanced by a heat denaturation process and the use of a detergent-supplemented trypsin solution (Fig. 2.7). For these reasons, on-tissue digestion is an attractive alternative method for detecting proteins that cannot be ionized by standard methods. Details regarding this method are described in Chap. 5.

2.3.2 Application to Protein Studies

IMS, with the choice of appropriate protocols, can provide unique and informative molecular distribution images that cannot be obtained via other imaging methods. One of the advantages of MS is that it can distinguish distinct chemical-structural variations of analytes by their mass. Because of this advantage, traditional MS research has been applied to the identification of posttranslational modifications in proteins [17]. IMS also enables the detection of protein molecular species in which subdomains contain different amino acids, in addition to the visualization of the distribution of these proteins on tissue sections. For example, Stoeckli et al. visualized the amyloid β molecular species [18, 19], which is generated by cleaving the amyloid precursor protein at a different cleavage site [20]. Figure 2.8 reveals the simultaneous visualization of the distribution of five amyloid β variants in tissue prepared from a mouse model of Alzheimer's disease [18, 19]. Although immunohistochemistry requires specific antibodies againts each variant, IMS has the advantage of being able to determine the distribution of these molecular variants all at once, via a simple protocol.

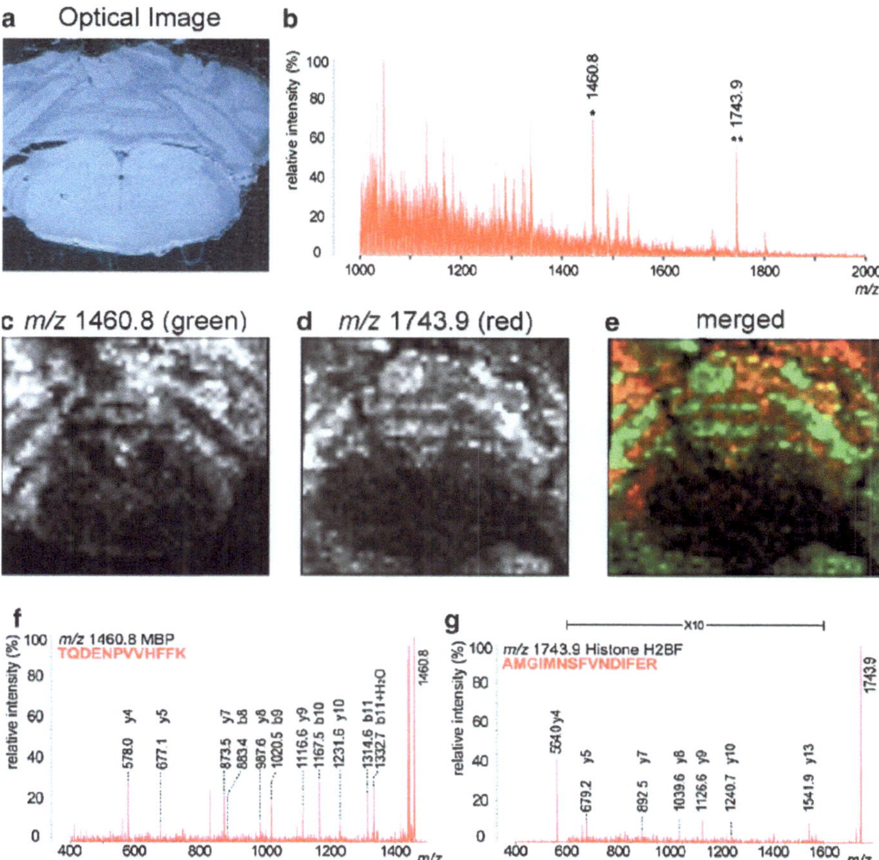

Fig. 2.6 Tryptic-digested protein imaging and precursor ion mass spectra with positive ion detection mode. (**a**) Optical image of imaging region and (**b**) accumulated mass spectrum from imaging region. (**c**, **d**) Imaging results of *m/z* 1,460.8 and 1,743.9, which are labeled by *asterisks* in (**b**). Merged image (*red*, *m/z* 1,460.8; *green*, *m/z* 1,743.9) is shown in (**e**). These peaks were identified by direct multistage tandem mass spectrometry (MSn) MSn as the fragment ions of myelin basic protein (MBP) (**f**) and histone H2B (**g**) (Reprinted from Shimma et al., Anal Chem 80(3):878–885.)

2.4 IMS for Small Organic Compounds

Today, applications of IMS to small organic compounds (i.e., $m/z < 1,000$) can be subdivided into two areas, as follows:

1. Measurement of <u>endogenous</u> small compounds (such as lipids)
2. Measurement of <u>exogenous</u> compounds (such as drugs)

In either application area, sample preparation strategies are different from those used for protein/peptide imaging. First, in general, tissue samples are not rinsed for the measurement of small compounds (Fig. 2.9), because small molecules tend to

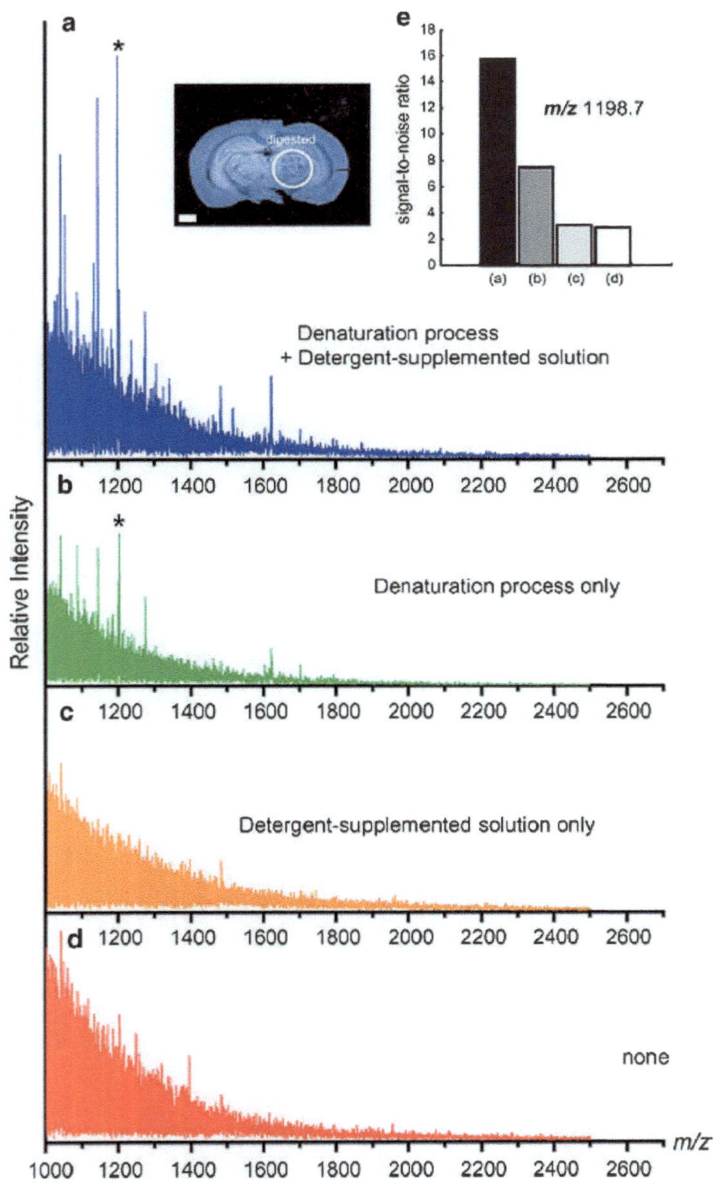

Fig. 2.7 On-tissue digestion process was enhanced by a heat denaturation process and the use of detergent-supplemented trypsin solution. Mass spectra were obtained from trypsin-digested protein from 10-μm sections of mouse cerebellum of (see *inset*). Tissue sections were prepared with the proposed protocol (denaturation process and detergent) (**a**), with denaturation and without detergent (**b**), without denaturation and with detergent (**c**), and with neither denaturation nor detergent (**d**). *Asterisks* represent the mass peaks at m/z 1,198.7. Signal-to-noise ratio of peak at m/z 1,198.7 was obtained from each tissue section (**e**). *Bar* 1 μm. (From [13]) (Reprinted from Setou et al., Applied Surface Sci 255(4):1555–1559.)

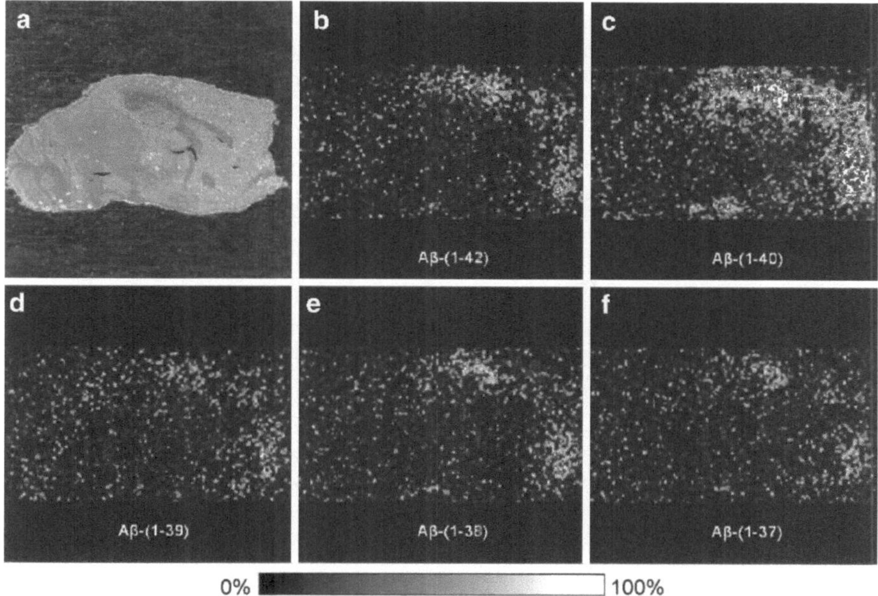

Fig. 2.8 Distribution of six amyloid β variants in the tissue prepared from a mouse model of Alzheimer's disease (AD). **a** Optical image of sagittal AD brain section; **b** Ab-(1–42) molecular image (*m/z* 4,515.1); **c** Ab-(1–40) molecular image (*m/z* 4,330.9); **d** Ab-(1–39) molecular image (*m/z* 4,231.7); **e** Ab-(1–38) molecular image (*m/z* 4,132.6); **f** Ab-(1–37) molecular image (*m/z* 4,075.5). (From [19]) (Reprinted from Rohner et al., Mech Ageing Dev 126:177–185.)

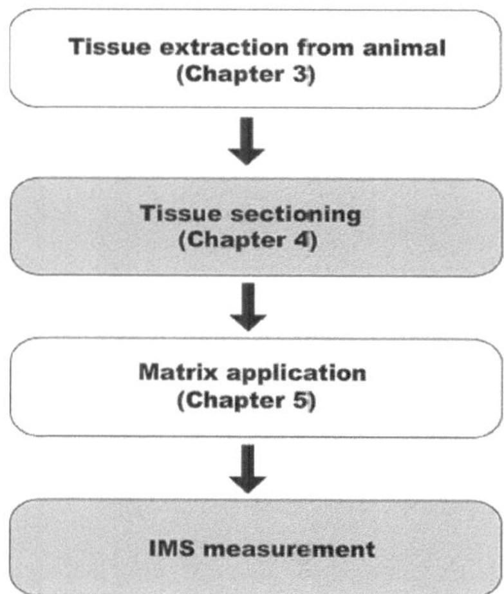

Fig. 2.9 A scheme for IMS of small molecules

be easily lost during the rinsing step, regardless of whether they are endogenous or exogenous. Even if they remain upon tissue sections, they might migrate within the tissue during rinsing, and thus a proper distribution image may not be obtained. Second, DHB or CHCA is often used as a matrix; along with these organic matrix compounds, the use of other ionization-enhancing reagents, such as nanoparticles [21, 22] and graphite [23, 24], has been recently developed. These methods have some advantages over the use of organic matrices, including the ability to eliminate matrix-derived peaks and improve spatial resolution [25]. Finally, by utilizing tandem MS analysis, the detailed structures of molecules can be identified, confirming whether observed mass signals are derived from the molecules of interest [16, 26].

Another approach for small molecular analysis is obtaining an image by scanning the tissue by tandem MS [26–28]. This approach provides structural information regarding parent ions of interest throughout the tissue, resulting in obtaining highly specific information on a target molecule. For example, this method could separate isomers that have roughly the same *m/z* value; Garrett et al. demonstrated that a peak at *m/z* 828 contains three kinds of ions – PC (diacyl-16:0/22:6), PS (diacyl-16:0/18:1), and DHB cluster – in rat brain by tandem MS scanning in a positive ion detection mode (Fig. 2.10) [26]. In addition to this, recently utilized ion-mobility separation adds a further separation step and provides more specific information [21, 28, 29].

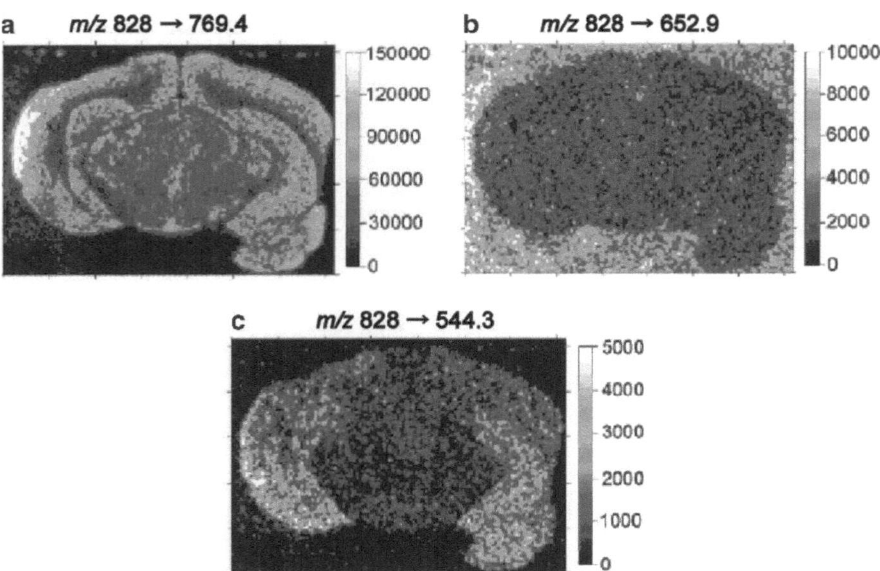

Fig. 2.10 An example of IMS by tandem MS, demonstrating that a peak at *m/z* 824 contains three kinds of ions: i.e., PC (diacyl-16:0/22:6), PS (diacyl-16:0/18:1), and DHB cluster. Each of these ion molecules can be visualized by tandem MS-IMS. (Partially modified from [26]) (Reprinted from Garrett et al., Int J Mass Spectrom 260(2-3):160–176.)

2.4.1 IMS for Endogenous Metabolites

Among endogenous compounds, lipid metabolites including phospholipids [4, 26, 30–32], neutral lipids (e.g., cholesterols) [31, 33], or glycolipids [24, 31, 34] have been well investigated. In addition, metabolite compounds with high UV absorbance for their cyclic structure (e.g., heme B) can also be ionized efficiently [4, 35, 36].

In planning the sample preparation strategies for these small molecules, there are several points to note, particularly in the analysis of endogenous phospholipids. Because the rinsing step is omitted, sodium and potassium salts remain on the tissue samples that generate sodium- and potassium-adducted phospholipid ions, in addition to the protonated ions. The generation of such multiple ion forms from a single analyte often perturbs otherwise successful IMS experiments. The distribution image of phospholipids might reflect no actual distribution of phospholipids, but rather heterogeneously distributed salts. Furthermore, because phospholipids comprise many molecular species with different fatty acids, a single peak might contain multiple kinds of ions. For example, a protonated PC (diacyl-16:0/20:4) molecule can be detected as having the same mass as a sodiated PC (diacyl-16:0/18:1) ion, at m/z 782. To overcome this problem, it is important to

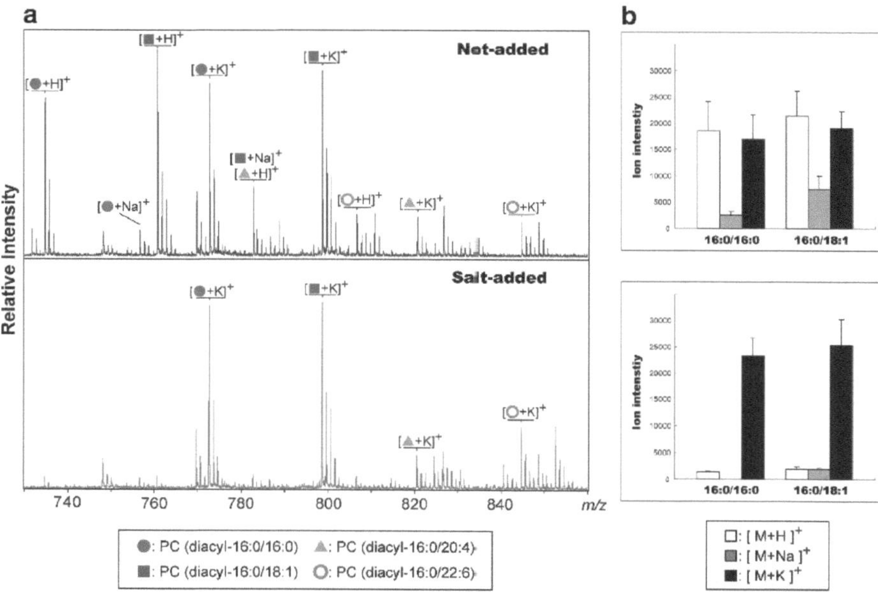

Fig. 2.11 Generation of multiple molecular ions from a single PC molecular species was reduced by adding an alkali-metal salt to the matrix solution. **a,** Spectra obtained from sections of mouse brain-homogenate using matrix solution with/without potassium acetate. The use of the salt-added matrix solution allowed multiple molecular ion-forms of PCs to be reduced to a single potassiated molecular ion form. **b,** Ion intensities of two PC species, PC (diacyl-16:0/16:0) and PC (diacyl-16:0/18:1), in three molecular ion forms, protonated, sodiated, and potassiated molucules [41,42]. Adapted from [41]

reduce the potential ion forms to a single alkali metal-adducted form. In this example, potassium salt could be added to the matrix solution (e.g., 20 mM potassium acetate, 40 mg ml^{-1} DHB, 70% MetOH). Through this procedure, the molecular ion forms would be limited to potassium-adducted molecules and the spectra would be simplified. As a result, the two types of PCs detected at m/z 782 could be separated at m/z 820 and m/z 798.

2.4.2   Application to Lipid Studies

As number of studies have shown, IMS is effective for profiling [4, 30, 37] and visualizing the distribution of phospholipids, especially on nerve tissue [21, 26, 29, 38]. A large content of phospholipids in the nerve cells and an easily charged structure yields a high ionization efficiency. In particular, in the positive ion mode, molecular species of choline-containing lipids (e.g., PC, lyso-PC, and sphingomyelin) are highly ionized because of their trimethylamine head group [7]. Phospholipid molecular species show localization within nerve tissues, depending upon the molecular species (i.e., fatty acid constituents). Figure 2.12 shows the distribution of five abundant phosphatidylcholines in a sagittal mouse brain section. Even in retinal tissue, which has a multilayered structure within thickness of approximately 150 μm, MALDI-MS revealed the distinct distribution of PC molecular species in a specific layer (Fig. 2.13) [38]. As currently only IMS can visualize such different phospholipid structures hidden in the cell membrane, the visualization of individual phospholipid molecular species will make great contributions to the field of lipidomics.

2.4.3   Imaging of Administered Drugs

There has been great interest surrounding the use of IMS as a tool for monitoring drug metabolism and/or drug delivery. For this purpose, there are many advantages to the use of IMS compared to whole-body autoradiography, which uses radiolabeled compounds. First, we can simultaneously monitor not only the parent drug but also their metabolites: the IMS ability to detect/visualize various molecules simultaneously allows researchers to image intact drugs separately from their metabolites. Second, IMS can visualize the distribution of drugs within a much shorter timeframe compared to detection using isotopes. The study groups of Caprioli and Stoeckli have made advances in this field [27, 39]. Figure 2.14 shows the detection of drugs that have been orally delivered in mice. Khatib-Shahidi et al. [27] showed the distribution of the drug olanzapine and its metabolite in a whole-rat sagittal section: a distinct distribution was seen 6 h after dosing. Although the intact drug reached the target organ (i.e., the brain), its metabolite localized in the bladder. Another advantage is that, by analyzing other endogenous metabolites, we are

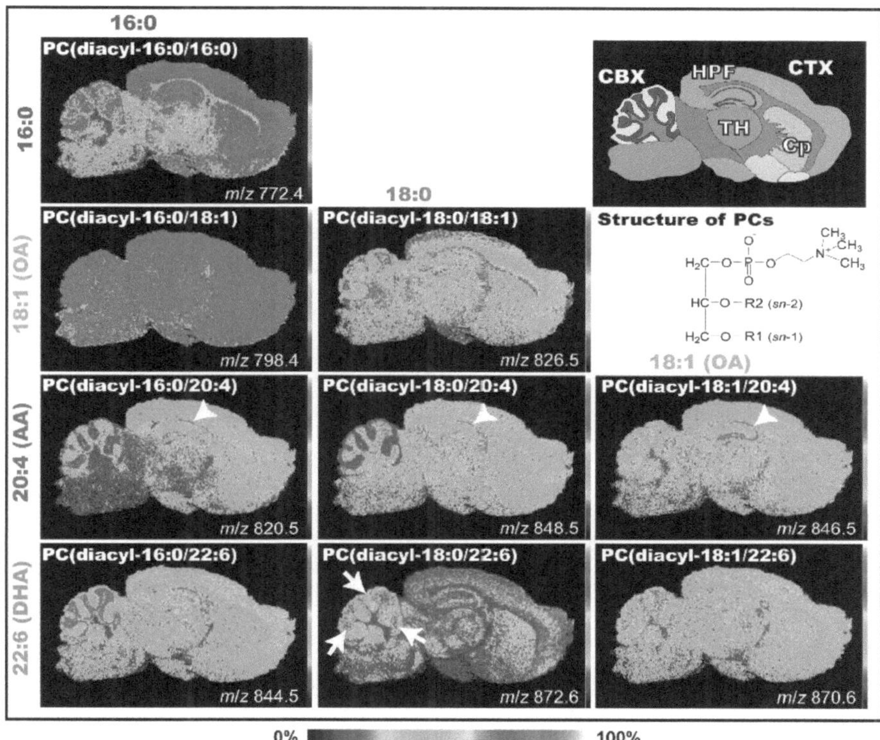

Fig. 2.12 Differential distribution of PC molecular species in sagittal mouse brain sections. MALDI-MS spectra of a brain section simultaneously identified the heterogeneous distributions of several PCs. Schema of the mouse brain sagittal section and ion images of PCs obtained by IMS are shown. Ion images of PCs are arranged according to their fatty acid (FA) composition. PCs with identical FA compositions at the sn-1 position are arranged lengthwise, while those with identical FA compositions at the sn-2 position are arranged sideways. The structures of PCs are also presented. Among the PCs, AA-PCs showed characteristic localization in the hippocampal cell layers (arrowheads). Among DHA-containing species, two abundant species, PC (diacyl-16:0/22:6) and PC (diacyl-18:1/22:6), were commonly enriched in the granule layer of the cerebellum, while PC (diacyl-18:0/22:6) showed a characteristic dotted distribution pattern near the cell layer (arrows).AA, arachidonic acid; DHA, docosahexaenoic acid; CBX, cerebellar cortex; CP, corpus striatum; CTX, cerebral cortex; HPF; hippocampal formation; TH, thalamus. Form literature [Visualization of the cell-selective distribution of PUFA-containing phosphatidylcholines. Sugiura Y, Konishi Y, Zaima N, Kajihara S, Nakanishi H, Taguchi R, Setou M., J. Lipid Res. 200p; 50:1776–1788]

potentially able to assess the drug responses of organs by analyzing other endogenous metabolites.

The adopted protocol for the sample preparation and selection of matrices is almost the same as that used in the analysis of endogenous small molecules. Because peaks of spectra in a small *m/z* region are very crowded by appearance

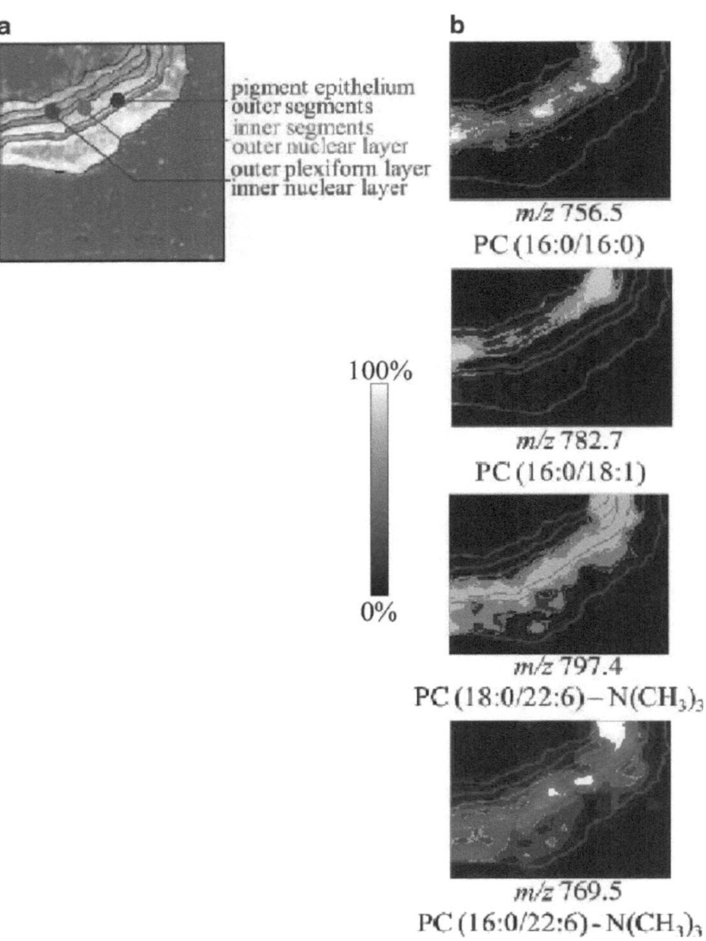

Fig. 2.13 Distribution of PC molecular species in a mouse retinal section. **a** Retina layer structure can be roughly distinguished in this optical image of the mouse retinal section, indicated as three colored dots. **b** The ion image merged from [PC (diacyl-16:0/16:0)+Na]⁺, [PC (diacyl-16:0/18:1)+Na]⁺, and [PC (18:0/22:+)+Na–N(CH₃)₃]⁺ revealed the three-zone distribution of the retinal section. [PC (16:0/22:6)+Na–N(CH₃)₃]⁺ was distributed in the same region as [PC (18:0/22:6)+Na–N(CH₃)₃]⁺ (Reprinted from Hayasaka et al., Rapid Commun Mass Spectrom 22:3415–3426.)

ofions of endogenous small molecules and matrix clusters, a mass peak for a particular drug might contain various ions. In such a case, the specific distribution image for drugs can be visualized by advanced MS imaging (e.g., imaging with MS² and MS³ mode, and ion mobility separation) as previously described (see Fig. 2.14).

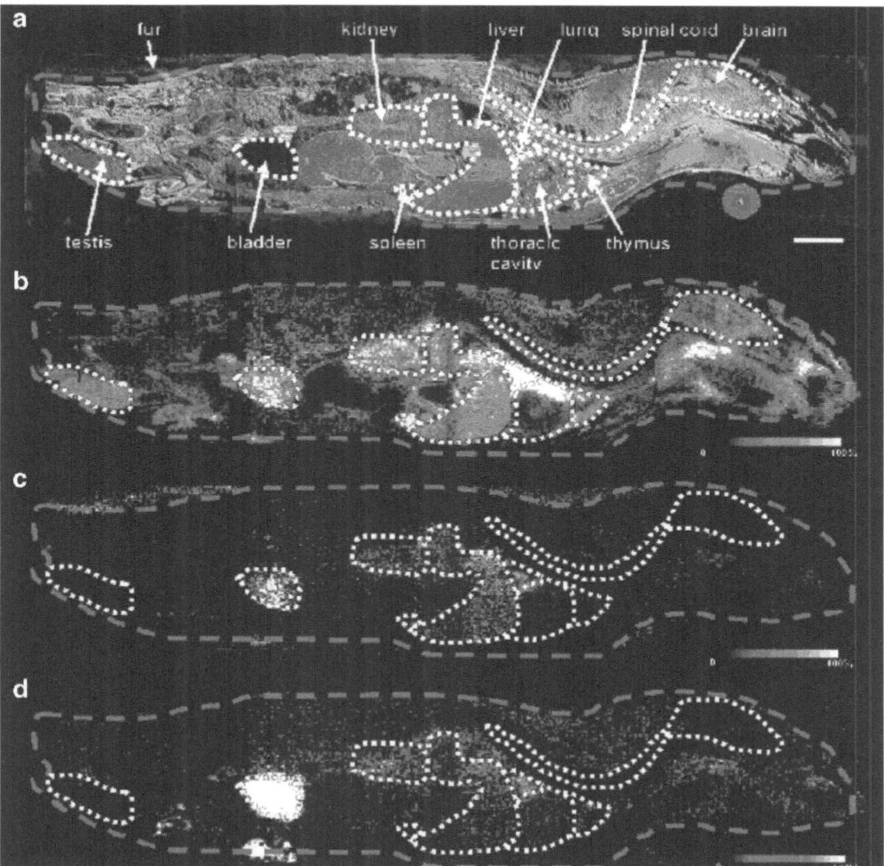

Fig. 2.14 Detection of drug and metabolite distribution in a whole-rat sagittal tissue section by a single IMS analysis. **a** Optical image of a rat tissue section 2 h post olanzapine dosing across four gold MALDI target plates. **b** MS/MS ion image of olanzapine (*m/z* 256). **c** MS/MS ion image of *N*-desmethyl metabolite (*m/z* 256). **d** MS/MS ion image of 2-hydroxymethyl metabolite (*m/z* 272). *Bar* 1 cm (Reprinted from Khatib-Shahidi et al., Anal Chem 78(18):6448–6456.)

One point to note in making drug measurements is that drugs might not be ionized equally in some parts of tissue regions. Indeed, by using a whole-mouse body section coated with drugs, Stoeckli et al. have found small regions (although albeit <5% of the total region) from which drugs could not be detected, presumably because of the ion suppression effect [39]. Although MALDI-MS has a disadvantage in that the objective peak can be affected by other molecules existing in the tissues, many such inherent problems can be avoided mitigated in preliminary experiments by using reference drug samples. Atkinson et al. provided an excellent example regarding this issue; they sought to visualize the distribution of a

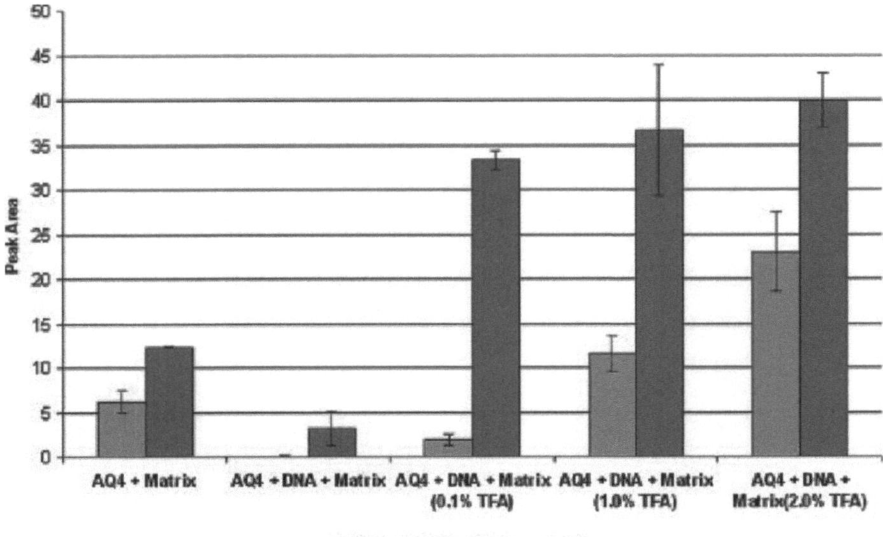

Fig. 2.15 Optimization of the condition of ionization for the drug AQ4. Because AQ4 forms complexes with DNA, a high concentration of trifluoroacetic acid (TFA) was added to avoid the complex formation. (From [40]) (Reprinted from Atkinson et al., Rapid Commun Mass Spectrom 21:1271–1276.)

compound, AQ4, and found that it is hardly ionized by itself because it forms complexes with DNA. In screening of composition of matrix solution they found appropriate condition that can support the ionization of AQ4 without generating complexes, by increasing acidity of matrix solution. Using this matrix, they succeeded in visualizing the distribution of AQ4 (Fig. 2.15) [40]. As in this case, depending upon the properties of the analyte, it is important to perform preliminary experiments to choose suitable matrices and obtain information regarding fragmentation patterns.

2.5 Conclusion

The importance of choosing an appropriate experimental protocol, particularly for the sample preparation step discussed here, is evident. As discussed, by providing some technical points MALDI-IMS provides valuable information with high selectivity, rapid acquisition, and the parallel acquisition of multiple analytes. We believe that improvements of experimental protocols will further expand the capabilities of this emerging technique. We hope that this chapter helps our readers explore a new, IMS-based frontier in their respective fields.

References

1. Krause E, Wenschuh H, Jungblut PR (1999) The dominance of arginine-containing peptides in MALDI-derived tryptic mass fingerprints of proteins. Anal Chem 71:4160–4165
2. Gharahdaghi F, Kirchner M, Fernandez J, et al. (1996) Peptide-mass profiles of polyvinylidene difluoride-bound proteins by matrix-assisted laser desorption/ionization time-of-flight mass spectrometry in the presence of nonionic detergents. Anal Biochem 233:94–99
3. Annesley TM (2003) Ion suppression in mass spectrometry. Clin Chem 49:1041–1044
4. Jones JJ, Borgmann S, Wilkins CL, et al. (2006) Characterizing the phospholipid profiles in mammalian tissues by MALDI FTMS. Anal Chem 78:3062–3071
5. Lemaire R, Wisztorski M, Desmons A, et al. (2006) MALDI-MS direct tissue analysis of proteins: improving signal sensitivity using organic treatments. Anal Chem 78:7145–7153
6. Li Y, Shrestha B, Vertes A (2007) Atmospheric pressure molecular imaging by infrared MALDI mass spectrometry. Anal Chem 79:523–532
7. Pulfer M, Murphy RC (2003) Electrospray mass spectrometry of phospholipids. Mass Spectrom Rev 22:332–364
8. Schwartz SA, Reyzer ML, Caprioli RM (2003) Direct tissue analysis using matrix-assisted laser desorption/ionization mass spectrometry: practical aspects of sample preparation. J Mass Spectrom 38:699–708
9. Seeley EH, Oppenheimer SR, Mi D, et al. (2008) Enhancement of protein sensitivity for MALDI imaging mass spectrometry after chemical treatment of tissue sections. J Am Soc Mass Spectrom 19:1069–1077
10. Kussmann M, Nordhoff E, Rahbek NH, et al. (1997) Matrix-assisted laser desorption/ionization mass spectrometry sample preparation techniques designed for various peptide and protein analytes. J Mass Spectrom 32:593–601
11. Altelaar AFM, Taban IM, McDonnell LA, et al. (2007) High-resolution MALDI imaging mass spectrometry allows localization of peptide distributions at cellular length scales in pituitary tissue sections. Int J Mass Spectrom 260:9
12. Katayama H, Nagasu T, Oda Y (2001) Improvement of in-gel digestion protocol for peptide mass fingerprinting by matrix-assisted laser desorption/ionization time-of-flight mass spectrometry. Rapid Commun Mass Spectrom 15:1416–1421
13. Setou M, Hayasaka T, Shimma S, et al. (2008) Protein denaturation improves enzymatic digestion efficiency for direct tissue analysis using mass spectrometry. Appl Surface Sci 255(4):1555–1559
14. Groseclose MR, Andersson M, Hardesty WM, et al. (2007) Identification of proteins directly from tissue: *in situ* tryptic digestions coupled with imaging mass spectrometry. J Mass Spectrom 42:254–262
15. Shimma S, Furuta M, Ichimura K, et al. (2006) A novel approach to *in situ* proteome analysis using chemical inkjet printing technology and MALDI-QIT-TOF tandem mass spectrometer. J Mass Spectrom Soc Jpn 54:133–140
16. Shimma S, Sugiura Y, Hayasaka T, et al. (2008) Mass imaging and identification of biomolecules with MALDI-QIT-TOF-based system. Anal Chem 80:878–885
17. Mann M, Jensen ON (2003) Proteomic analysis of post-translational modifications. Nat Biotechnol 21:255–261
18. Yan R, Bienkowski MJ, Shuck ME, et al. (1999) Membrane-anchored aspartyl protease with Alzheimer's disease beta-secretase activity. Nature (Lond) 402:533–537
19. Rohner TC, Staab D, Stoeckli M (2005) MALDI mass spectrometric imaging of biological tissue sections. Mech Ageing Dev 126:177–185
20. Stoeckli M, Staab D, Staufenbiel M, et al. (2002) Molecular imaging of amyloid beta peptides in mouse brain sections using mass spectrometry. Anal Biochem 311:33–39
21. Jackson SN, Ugarov M, Egan T, et al. (2007) MALDI-ion mobility-TOFMS imaging of lipids in rat brain tissue. J Mass Spectrom 42:1093–1098
22. Moritake S, Taira S, Sugiura Y, et al. (2008) J Nanosci Nanotechnol (special issue) Adv Mat Nanosci Nanotechnol 8: (in press)

23. Zhang H, Cha S, Yeung ES (2007) Colloidal graphite-assisted laser desorption/ionization MS and MS(n) of small molecules. 2. Direct profiling and MS imaging of small metabolites from fruits. Anal Chem 79:6575–6584
24. Cha S, Yeung, ES (2007) Colloidal graphite-assisted laser desorption/ionization mass spectrometry and MSn of small molecules. 1. Imaging of cerebrosides directly from rat brain tissue. Anal Chem 79:2373–2385
25. Taira S, Sugiura Y, Moritake S, et al. (2008) Nanoparticle-assisted laser desorption/ionization based mass imaging with cellular resolution. Anal Chem 80:4761–4766
26. Garrett TJ, Prieto-Conaway MC, Kovtoun V, et al. (2006) Imaging of small molecules in tissue sections with a new intermediate-pressure MALDI linear ion trap mass spectrometer. Int J Mass Spectrom 260:11
27. Khatib-Shahidi S, Andersson M, Herman JL, et al. (2006) Direct molecular analysis of whole-body animal tissue sections by imaging MALDI mass spectrometry. Anal Chem 78:6448–6456
28. Trim PJ, Henson CM, Avery JL, et al. (2008) Matrix-assisted laser desorption/ionization-ion mobility separation-mass spectrometry imaging of vinblastine in whole body tissue sections. Anal Chem 80:8628–8634
29. McLean JA, Ridenour WB, Caprioli RM (2007) Profiling and imaging of tissues by imaging ion mobility-mass spectrometry. J Mass Spectrom 42:1099–1105
30. Rujoi M, Estrada R, Yappert MC (2004) *In situ* MALDI-TOF MS regional analysis of neutral phospholipids in lens tissue. Anal Chem 76:1657–1663
31. Jackson SN, Wang HY, Woods AS (2005) Direct profiling of lipid distribution in brain tissue using MALDI-TOFMS. Anal Chem 77:4523–4527
32. Puolitaival SM, Burnum KE, Cornett DS, et al. (2008) Solvent-free matrix dry-coating for MALDI imaging of phospholipids. J Am Soc Mass Spectrom 19:882–886
33. Altelaar AF, Klinkert I, Jalink K, et al. (2006) Gold-enhanced biomolecular surface imaging of cells and tissue by SIMS and MALDI mass spectrometry. Anal Chem 78:734–742
34. Sugiura Y, Shimma S, Konishi Y, et al. (2008) Imaging mass spectrometry technology and application on ganglioside study: visualization of age-dependent accumulation of C20-ganglioside molecular species in the mouse hippocampus. PLoS ONE 3:e3232
35. Shimma S, Setou M (2007) Mass microscopy to reveal distinct localization of heme B (*m/z* 616) in colon cancer liver metastasis. J Mass Spectrom Soc Jpn 55:145–148
36. Mazel V, Richardin P, Debois D, et al. (2007) Identification of ritual blood in African artifacts using TOF-SIMS and synchrotron radiation microspectroscopies. Anal Chem 79:9253–9260
37. Jackson SN, Wang HY, Woods AS (2005) *In situ* structural characterization of phosphatidylcholines in brain tissue using MALDI-MS/MS. J Am Soc Mass Spectrom 16:2052–2056
38. Hayasaka T, Goto-Inoue N, Sugiura Y, et al. (2008) Matrix-assisted laser desorption/ionization quadrupole ion trap time-of-flight (MALDI-QIT-TOF)-based imaging mass spectrometry reveals a layered distribution of phospholipid molecular species in the mouse retina. Rapid Commun Mass Spectrom 22:3415–3426
39. Stoeckli M, Staab D, Schweitzer A (2006) Compound and metabolite distribution measured by MALDI mass spectrometric imaging in whole-body tissue sections Int J Mass Spectrom 260(2-3):195–202
40. Atkinson SJ, Loadman PM, Sutton C, et al. (2007) Examination of the distribution of the bioreductive drug AQ4N and its active metabolite AQ4 in solid tumours by imaging matrix-assisted laser desorption/ionisation mass spectrometry. Rapid Commun Mass Spectrom 21:1271–1276
41. Sugiura Y, Konishi Y, Zaima N, Kajihara S, Nakanishi H, Taguchi R, and Setou M (2009) Visualization of the cell-selective distribution of PUFA-containing phosphatidylcholines in mouse brain by imaging mass spectrometry. J Lipid Res 50:1776–1788
42. Sugiura Y, Setou M. (2009) Selective imaging of positively charged polar and nonpolar lipids by optimizing matrix solution composition, Rapid Commun Mass Spectrom, 30;23(20):3269–78

Part III
Sample Preparation

Chapter 3
Animal Care and Tissue Sample Extraction for IMS

Ikuko Yao and Mitsutoshi Setou

Abstract In this chapter, we describe the sampling of biological materials by using an example of the removal of the brain from a mouse and its cryopreservation. When we use a biomaterial as an IMS sample, one aspect that must be taken into account is an accurate reflection of the ante mortem conditions of the living being as far as possible by extracting the sample as soon as the animal dies. We describe the methods often used for mice regarding euthanasia and removal and freezing of the mouse brain.

3.1 Methods of Mouse Euthanasia

Not only lethal actions but also other treatment of laboratory animals should be based on laws and ordinances and the policies of jurisdictive organization. When lethal action toward the animal is needed, pain should be avoided; loss of consciousness followed by quick death is ideal without exciting the animal with undue restraint, etc. (avoidance of pain, excitement, fear, insecurity, depressed state, etc.). Further, experimenters who must carry out lethal actions always ask for secure and simple treatments. Next, we describe the methods often used for sacrifice of mice.

3.1.1 Euthanasia by Anesthetics

An overdose of pentobarbital causes a quick, painless death without anxiety and agitation. Intravenous administration may be sufficient, although intraperitoneal injection is used in many cases. The dose required for euthanasia is approximately four- to fivefold times the dose used for anesthesia. Other anesthetic drugs, such as

I. Yao
Department of Medical Chemistry, Kansai Medical University, Fumizono-cho 10-15, Moriguchi, Osaka, Japan

M. Setou (✉)
Department of Molecular Anatomy, Hamamatsu University School of Medicine, 1-20-1 Handayama, Higashi-ku, Hamamatsu, Shizuoka 431-3192, Japan
e-mail: setou@hama-med.ac.jp

M. Setou (ed.), *Imaging Mass Spectrometry: Protocols for Mass Microscopy*,
DOI 10.1007/978-4-431-09425-8_3, © Springer 2010

other barbitals, ketamine, and xylazine mixtures, may be used. Ether may also be used. As ether is explosive in nature, ether volatilization and dispersion should be performed before waste disposal.

3.1.2 Cervical Dislocation

Cervical dislocation by a skilled experimenter can bring death to an animal without sharp pain, insecurity, and excitement even without anesthesia. By holding the head of a mouse on the left and right with the thumb and index finger, and pulling the tail (or the hind limbs) quickly and firmly, the intracranial brain tissues are made to dissociate from the cervical spinal cord at the position of the first cervical vertebra, following which the animal's awareness is nullified. It is assumed that respiratory arrest will occur immediately.

3.1.3 Decapitation

In pharmacology or in a biochemical experiment, when it is necessary to avoid the effect of pharmacological hypoxia, decapitation by guillotine is performed. If performed skillfully, the time for which the mouse is physically restrained is shortened, and death causes neither uneasiness nor pain. Because a mouse can struggle and inflict an injury on the operator, while also lengthening the duration of its physical restraints, laboratory leaders need to teach this technique with responsibility. In addition, for a mouse, fatality caused by bloodletting, electricity, or a blow to the head is banned under law.

3.2 Anatomy of the Mouse

Next, we describe procedures for removal of the mouse brain:

1. Cut the heads off mice and turn over the skin to the face side as follows. Cut the back of the occipital foramen and parts of the atlas with a scissors. Ruffle through the skin at the parietal regions and temporal regions, from the nuchal end of the mouse toward the rostral dorsal and occipital bones (Fig. 3.1).
2. In the parietal region, insert the scissor straight from the foramen magnum and break the bone up to the bregma. Use a small scissors so as not to damage the brain surface at this time (Fig. 3.2).
3. Turn over and remove the skull from the cleavage sites of the parietal bone with the scissors. Insert the other edge into the gap between the skull and brain while opening the edge of the scissors a little, and remove according to the principle of the lever. Similarly, open both sides like gates (Fig. 3.3, upper panel).
4. Cut the edge of the nasal bones with scissors, and further open the frontal bones and the roots of the nasal bones from the center to the right and left (Fig. 3.3, middle panel).

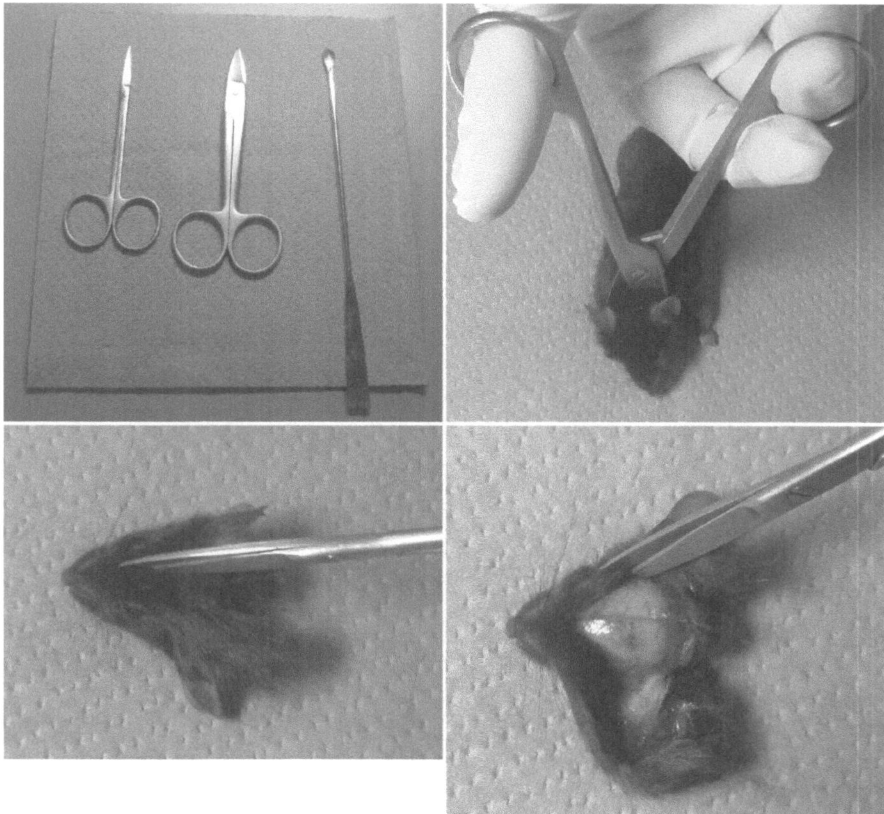

Fig. 3.1 Evisceration of mouse brain and decapitation: ruffling through mouse skin

5. Bring out the brain using a spatula. Remove the brain from the bottom by the
   side of the olfactory nerve or the cerebellum, so that it is not damaged. The
   removed brain should be fixed by freezing immediately (Fig. 3.3, bottom
   panel).

3.3   Freezing of the Mouse Brain

3.3.1   Materials

- Dry ice
- Cotton work gloves
- Wooden hammers
- Strong bags (bag made of cloth sail, etc.)
- Sieve

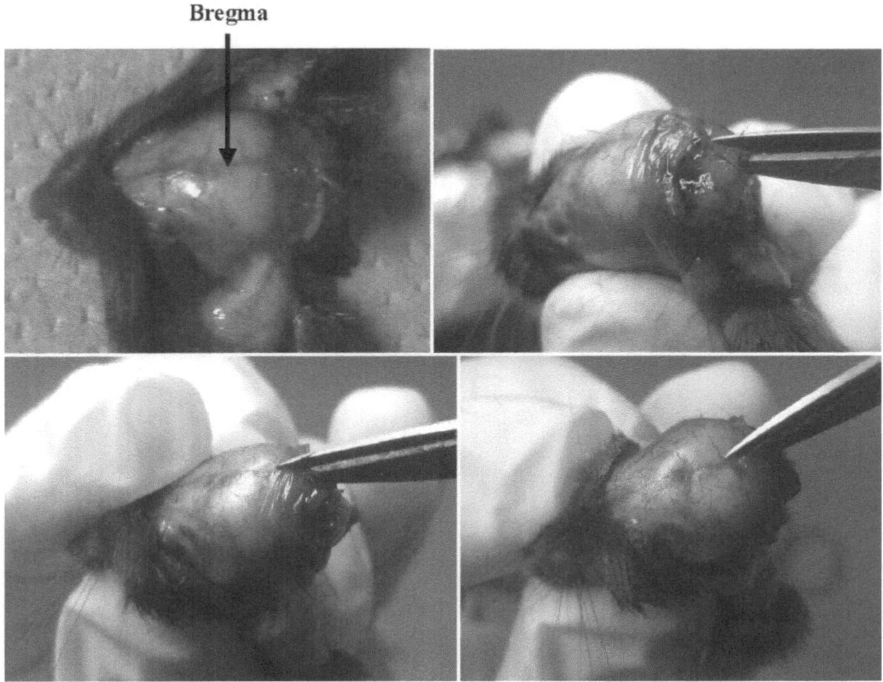

Fig. 3.2 Evisceration of mouse brain: cutting the skull bones of the mouse

3.3.2 Procedure

1. Break up the dry ice into pieces with a wooden hammer and then powder it as much as possible. Place the broken dry ice in the sieve and separate the powder snow from the large blocks (Fig. 3.4).
2. Embed the removed brain immediately in the powdered dry ice and freeze it quickly.
3. After checking that it is frozen completely, after approximately 1 min, place the brain into a plastic bag with a zipper, labeled with the sample name, and store in the deep freezer at −80°C until slice preparation (see Fig. 3.4).

To reflect the ante mortem condition of the animal, it is desirable to freeze the sample as quickly as possible. Thus, the magnitude of the ice crystals formed within a cell can be reduced, and membrane breakdown caused by freezing and thawing can be stopped or minimized. Although liquid nitrogen is often used for quick freezing, it is not recommended in this case because the tissue surface freezes so quickly that the tissue block will be broken by distortions arising from the differences between the expansion coefficients of the surface and the internal tissue.

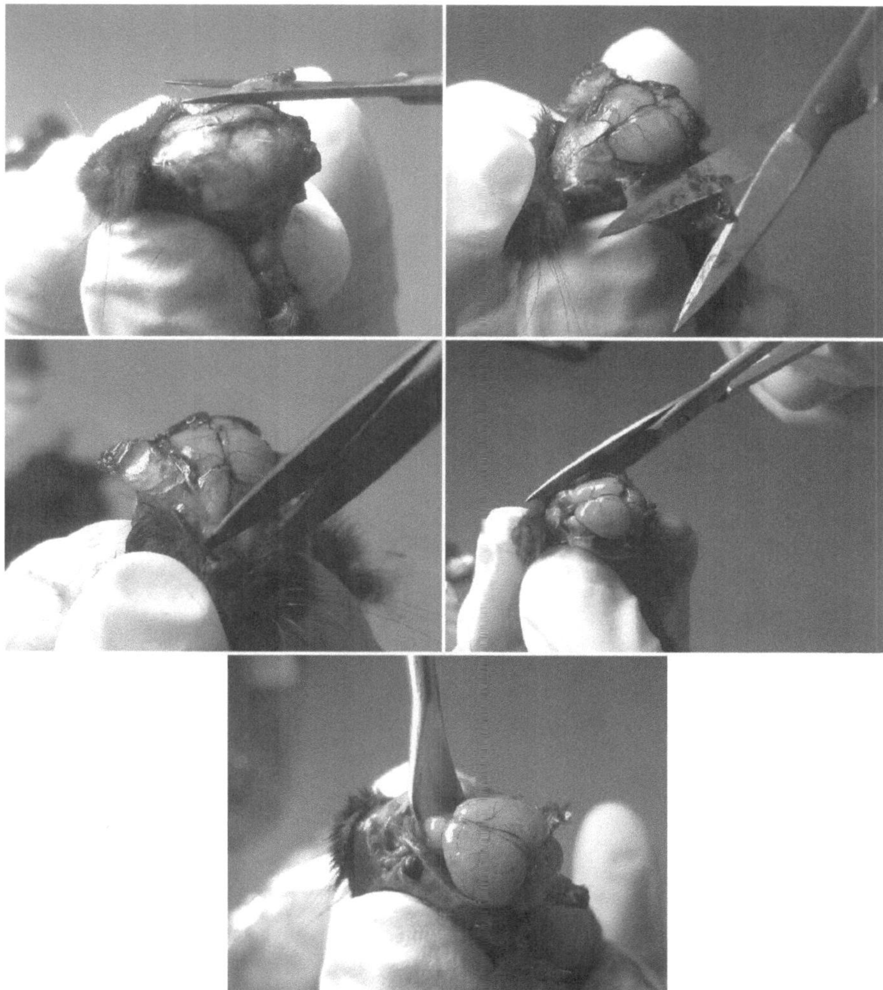

Fig. 3.3 Evisceration of mouse brain, bringing out the brain using a spatula

We generally use a freezing method with powdered dry ice because this method enables us to preserve samples in good condition that are frozen equally overall.

3.4 Samples Other Than Mouse Brain

Powdered dry ice should be used for freezing other tissues as well. Although immersion fixation is generally used in the case of postmortem human samples, the frozen sample is more preferable, because a favorable spectrum cannot be easily obtained by the bridge construction of protein molecules by formalin fixation.

Fig. 3.4 Manufacturing powdered dry ice

3.5 Evaluation of the Time Elapsed After Death in Animals

The phenomena that occur in the body immediately postmortem, including physical, biological, and chemical changes, are called "postmortem phenomena or postmortem appearance." These alterations begin immediately after the death of the animal. In particular, brain tissue is extremely susceptible to damage by proteases causing autolysis. Therefore, it is better to use fresh materials to obtain repeatable results. If you use a corpse as an experimental material by necessity, you should consider the postmortem phenomena as follows. Postmortem phenomena are important factors in calculating the time elapsed since death, and we often obtain useful information from them for determining the cause and manner of death. However, we have to be aware that actual after-death lapsed time varies because the processes of these postmortem appearances depend on the environment where the body is set and, greatly, on individual differences.

In the early time period, postmortem lividity, rigor mortis, fall in temperature, and muddiness or dryness of the cornea occur; if further time elapses, autolysis and corruption will appear as the main phenomena. Although these changes do not happen in laboratory animals, under certain environments, decomposition may also take place.

Chapter 4
Preparing Biological Tissue Sections for Imaging Mass Spectrometry

Yuki Sugiura and Mitsutoshi Setou

Abstract The process of making a thin slice section from frozen tissue blocks to the matrix application is described. Because the slice section created is served to direct mass spectrometry, there are some pivotal differences from other staining methods such as immunohistochemistry. In this chapter, the problems specific to mass spectrometry (MS) are introduced, and thereafter the ways of dealing with these problems are explained; last, the standard protocols are introduced.

4.1 Introduction

This section is devoted to the preparation of tissue sections for imaging mass spectrometry (IMS) i.e., the process of preparing thinly sliced sections from freshly frozen blocks of tissue to where these are served to the matrix application. The process of preparing sections for IMS measurement is essentially similar to that used in the preparation of frozen sections for immunostaining or dye-staining. However, because the sections created in this case are served to MS measurement, there are certain essential differences from other staining methods. A point to bear in mind is that IMS is an imaging technique based on MS, a microchemical analytical technique. In the preparation of such tissue sections, then, the use of an optimal cutting temperature compound (OTC) should be avoided, because contamination with such polymer

Y. Sugiura
Department of Bioscience and Biotechnology, Tokyo Institute of Technology,
4259 Nagatsuta-cho, Midori-ku, Yokohama, Kanagawa 226-8501, Japan

Y. Sugiura and M. Setou (✉)
Department of Molecular Anatomy, Hamamatsu University School of Medicine,
1-20-1 Handayama, Higashi-ku, Hamamatsu, Shizuoka 431-3192, Japan
e-mail: setou@hama-med.ac.jp

M. Setou (ed.), *Imaging Mass Spectrometry: Protocols for Mass Microscopy*,
DOI 10.1007/978-4-431-09425-8_4, © Springer 2010

Table 4.1 Major cautionary points at each step of slice preparation

Procedure for preparing a frozen section for IMS

Tissue extraction

Because of rapid metabolic turnover, tissues should be handled in a fixed time course, particularly for the analysis of small metabolite molecules

↓

Embedding

Avoid the use of polymeric compounds, such as optimal cutting temperature compound (OTC), for embedding [1, 2]

As an alternative, a precooled semiliquid gel of 2% sodium carboxymethylcellulose (CMC) can be used as an embedding compound, as it does not interfere with MS [3]

↓

Sectioning

Tissue thickness <20 μm improves spectrum quality especially for protein and peptide [4]

↓

Mounting

The use of conductive materials is recommended for supporting the tissue section [5][a]

↓

Washing

By removing small molecules such as lipids, the tissue-washing process improves the spectrum quality for protein/peptide detection [1, 6, 7]

↓

Drying

Failure to dry sections sufficiently may cause samples to peel off in the vacuum chamber of the mass spectrometer

↓

Sputtering

Sputtering of metals (e.g., gold) onto tissue sections before/after the matrix application process improves the spectrum quality [3, 8]

[a]Some instruments support the use of nonconductive materials [9]

molecules in MS causes a serious supression of biomolecule ionization [1]. The preclusion of OTC is a unique feature of the preparation protocol that distinguishes it from those used with immunohistochemical staining and other dye-staining methods. Table 4.1 summarizes such key points at each step in the section preparation process. In later paragraphs, problems specific to MS are introduced, as are techniques for handling these problems.

4.2   Embedding

In general, thin slices are prepared from a freshly frozen tissue block which has been penetrated with an embedding agent, such as the OTC and undergone freezing fixation. The tissues in the embedding agent are then cut into very thin slices. Embedding allows the sample to retain a good shape and facilitates the cutting process. However, in IMS experiments, it is known that the attachment and penetration of embedding agents (such as OTCs) in samples leads to a deterioration in the MS signals that are obtained [1, 2]. Such polymer-like resin compounds are commonly used as embedding agents and, in MS, these compounds have a high ionization efficiency; this leads to a decrease in the detection sensitivity of other molecules that would otherwise be observed. In particular, at analysis of small molecules with an m/z value of 1,000–2,000, contamination with OTC would lead to the emergence of extremely high polymer peaks in the mass spectra of positive ions, peaks that would virtually hide all the other smaller peaks (Fig. 4.1).

It also cause a decrease in the sensitivity of the signals when detecting higher molecular weight proteins [1, 2]. For all these reasons, OTC is used only for "supporting" the tissue blocks (as shown in Fig. 4.2a), such that it does not directly attach to the tissue sections. However, in the absence of the embedding process, difficulties may be encountered in cutting certain tissues into thin slice sections. Avoiding this problem, Stockli et al. used a precooled semiliquid gel of 2% sodium carboxymethylcellulose (CMC) as an alternative embedding compound that does not interfere with MS [3].

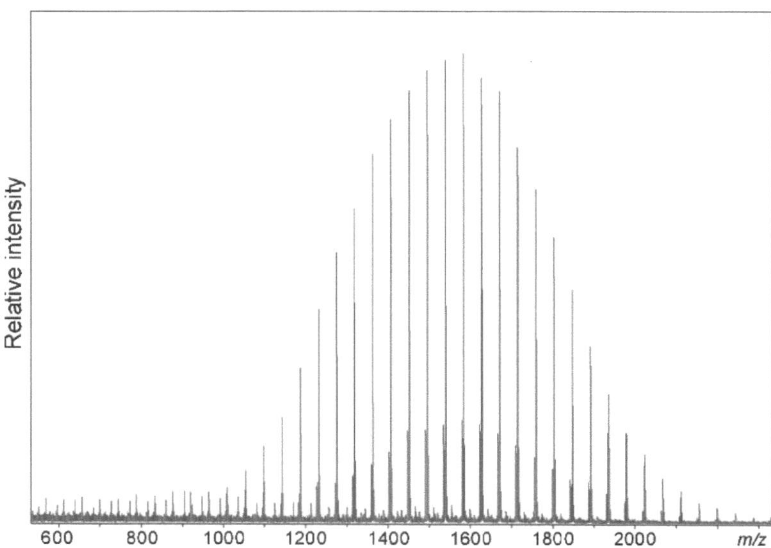

Fig. 4.1 Mass spectra of small molecules were obtained from a representative section that was completely embedded in optimal cutting temperature compound (OTC)

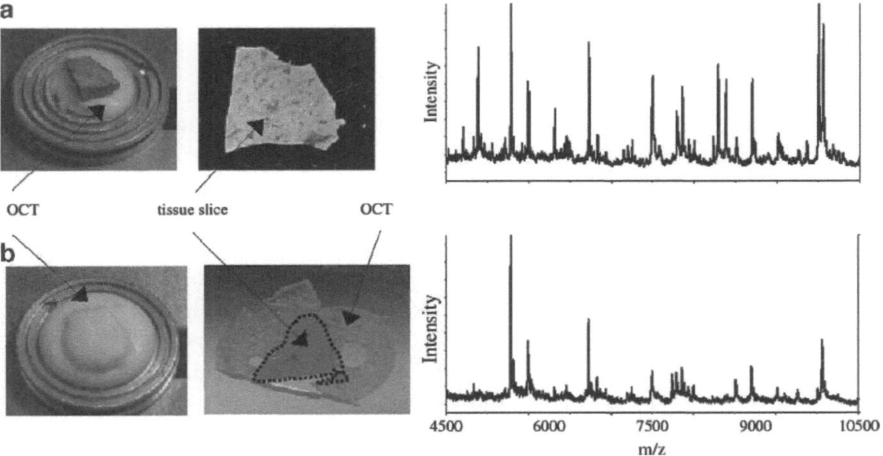

Fig. 4.2 Decrease in detection sensitivity of the ions originating from proteins because of contamination with OTC. OTC adhering to the tissue section diminishes the detectable peaks. (**a**) A case in which OTC was used only for supporting the tissue block. (**b**) A case in which the tissue block was completely embedded with OTC. (From [1]). (Reprinted from Schwartz et al., J Mass Spectrom 38:699–708.)

4.3   Excision of Thin Slices

Another important factor associated with the excision of thin slices for IMS is slice thickness. When the thickness of a slice is >15 μm, there is a deterioration in the detection sensitivity of IMS, particularly when analyzing high molecular weight proteins [4]. Figure 4.3 shows the mass spectra obtained from the cerebral cortex region in mouse brain slices with thicknesses of 2, 5, 10, 15, and 30 μm. A larger number of mass peaks with high signal-to-noise (S/N) ratios were observed in the spectra obtained from sections with 2-, 5-, and 10-μm thicknesses than in those obtained from sections with 15- and 30-μm thicknesses. This difference can be attributed to a phenomenon referred to as the "charging effect" [10]. Generally, biological tissue sections have low intrinsic electric conductivity, and this tendency is considered more apparent with thicker sections than with thinner ones. Therefore, on the surface of a thick slice, the surface of the sample is kept insulated from the support stage, which in turn has an electric conductivity. In this state, a surplus electric charge generated by laser irradiation is not lost through the sample stage, thus generating multiple charged ions on the sample; this ultimately leads to a significant loss of sample ions that would otherwise reach the detector [10]. However, in the case of frozen sections prepared without fixation, a high technical proficiency is required to stably create slices with thicknesses of several micrometers each. Currently, most samples are prepared with a slice thickness of 10–20 μm [11, 12]; these medium-thick sections appear to provide a good compromise between optimal IMS performance and experi-

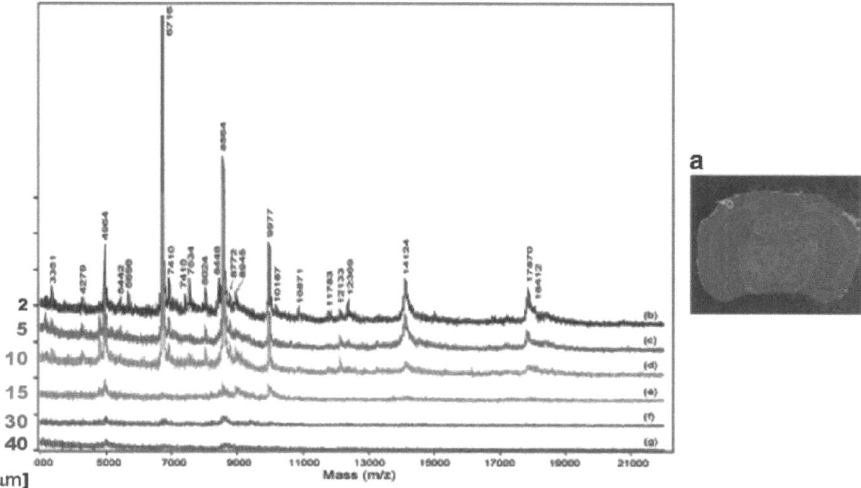

Fig. 4.3 Mass spectra obtained from the cerebral cortex region (see inset *a*) of mouse brain. Signals with good S/N ratios were obtained from thin slices. (From [1], modified). (Reprinted from Sugiura et al., J. Mass Spectrom Soc Jpn 54:45–48.)

mental efficiency [13], particularly when a large number of samples need to be analyzed [14]. The "metal sputtering" technique enhances the signal intensity and thus image quality [3, 8], presumably by avoiding the "charging effect."

4.4   Section Support Materials

Care should also be exercised in the selection of section support materials, i.e., the materials on which the tissue sections are laid. In IMS, in principle, support materials of which their surfaces are coated with conductive materials are used. In particular, when a tandem time-of-flight (TOF/TOF)-type instrument is used a high voltage is applied to the sample stage; therefore, omitting the use of conductive materials can lead to a significant reduction in detection sensitivity as a result of the aforementioned "charging effect." Exceptionally, a certain ion trap-type instrument makes it possible to measure the sample even on nonconductive materials [9].

The simplest technique involves placing the tissue slices directly onto a metal target plate [1]; however, in this case, when multiple measurements are made, the target plate must be thoroughly cleaned after each measurement. Currently, the method commonly used involves preparing samples on a disposable plastic sheet or a glass slide that is coated with a series of conductive materials [5]. Plastic sheets (ITO sheets) [15] or glass slides [5] (ITO glass slides; available from Bruker Daltonics or Sigma), coated with indium tin oxide (ITO), are particularly useful because they have a superior optical transparency that is suitable for microscopic

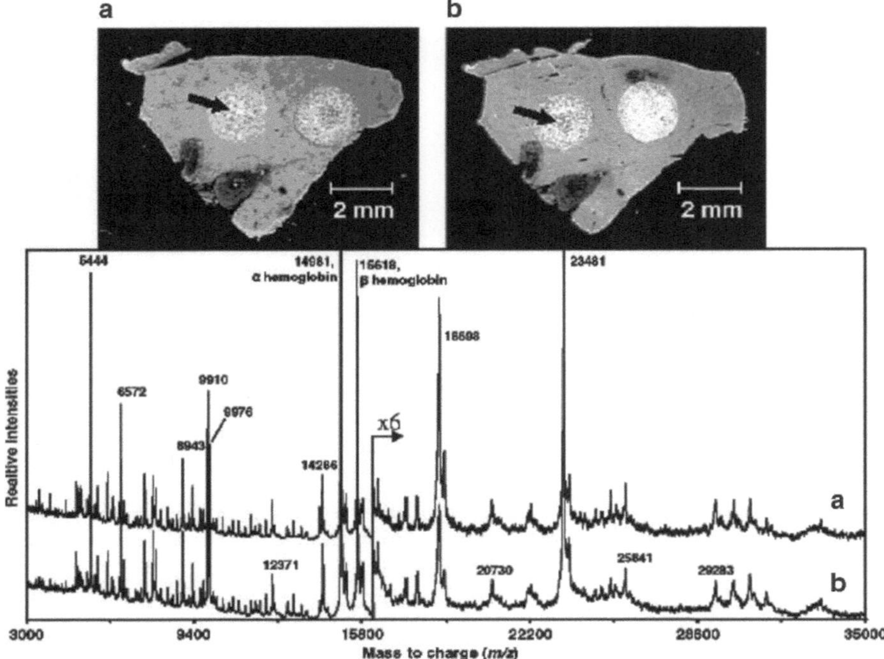

Fig. 4.4 Mouse liver sections (*upper panel*) on glass slides coated with gold (**a**) and ITO (**b**), and the corresponding mass spectra (*lower panel*). (Reprinted from Chaurand et al., Anal Chem 76(4):1145–1155.)

observations. In addition, they provide mass spectra with a high quality that is comparable to that obtained with materials coated with gold – the most conductive material (Fig. 4.4) [1, 5, 15]. ITO sheets are sufficiently thin (i.e., 125 μm) to be cut into appropriate shapes with scissors, and they also provide good light transmission [~85% visible light $(500 < \lambda < 900$ nm)]. In addition, the technique does not require the use of a specific adapter for introduction of glass slides into the mass spectrometer. Moreover, these materials are also versatile in that they can be attached to existing metal plates with conductive double-sided tape, which can then be introduced into any mass spectrometer [15, 16].

4.5   Postmortem Degradation

Another issue is the postmortem degradation of analyte molecules. In the sample preparation process, care needs to be exercised during both the organ extraction and section dehydration stages [17–19]. In both stages, a time-dependent appearance and a loss of signals may be observed [19], indicating that particular care is needed to ensure good sample-to-sample reproducibility.

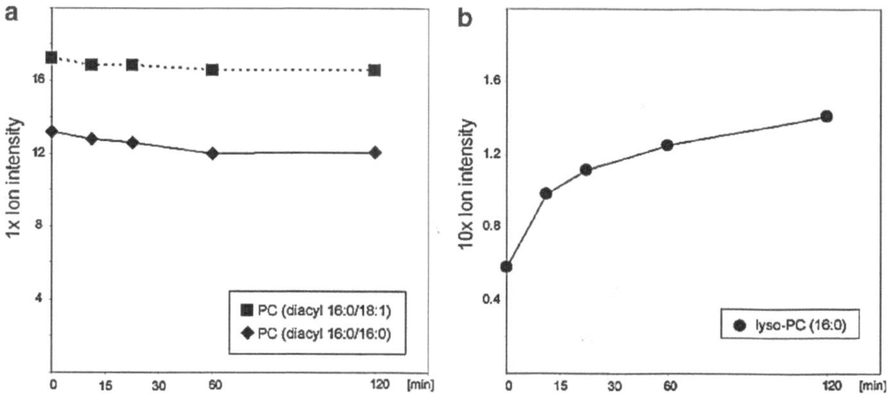

Fig. 4.5 Postmortem degradation of phosphatidylcholines and increase in lyso-PCs. We performed IMS on a series of mouse brains extracted at different time points after the mice had been sacrificed (<1, 15, 30, 60, and 120 min). After IMS, the ion intensities of the PCs were averaged over the entire section. As postmortem events, the degradation of PCs and an increase in lyso-PCs were observed within 15 min. These events were presumably caused by stimulation of the activity of phospholipases under ischemic conditions (Rehncrona et al. [20], Umemura et al. [21]). In this study, mouse brains were extracted within a minute (typically in 40 s) after sacrifice. (Reprinted from Sugiura et al., J Lipid Res 50:1776–1788.)

The first point is the time taken to extract organs. This is a critical parameter, particularly for the analysis of small molecules, owing to their rapid metabolic turnover [17, 18]. In our study involving an series of mouse brains extracted at different times after death (within 1, 15, 30, and 120 min), the postmortem degradation of phospholipids was observed within 15 min (Fig. 4.5). As postmortem events, the degradation of phospholipids and an increase in lyso-phospholipids were observed within 15 min, and they were presumably caused by a stimulation of the activity of phospholipases under ischemic conditions [20, 21]. The same applies to other metabolic molecules, such as dopamine and serotonin, which are metabolized into different molecules within a very short time period (within a few minutes) [17, 18]. All these factors suggest that the extraction of tissues should be performed as rapidly as possible. As an alternative method, the use of microwave heating devices is a relatively effective means of inactivating endogenous enzymes, thus eliminating postmortem degradation [22].

Second, the dehydration of sections before rinsing with organic solvents can also lead to changes in the signals that would otherwise be obtained [19]. Goodwin et al. experienced the time-dependent appearance and loss of signals when frozen tissue slices were brought rapidly to room temperature over a period of 30 s to 3 h. Sections of mouse brain were cut on a cryostat microtome, placed on a MALDI target, and allowed to warm to room temperature for the indicated period. Even within 1 min, protein/peptide signals were altered, in both an increasing and decreasing manner. Figure 4.6 shows three replicate experiments and time-courses of 0, 1, 2, 3, 4, 5, 6, and 7 min at room temperature; it demonstrates the typical changes observed over the

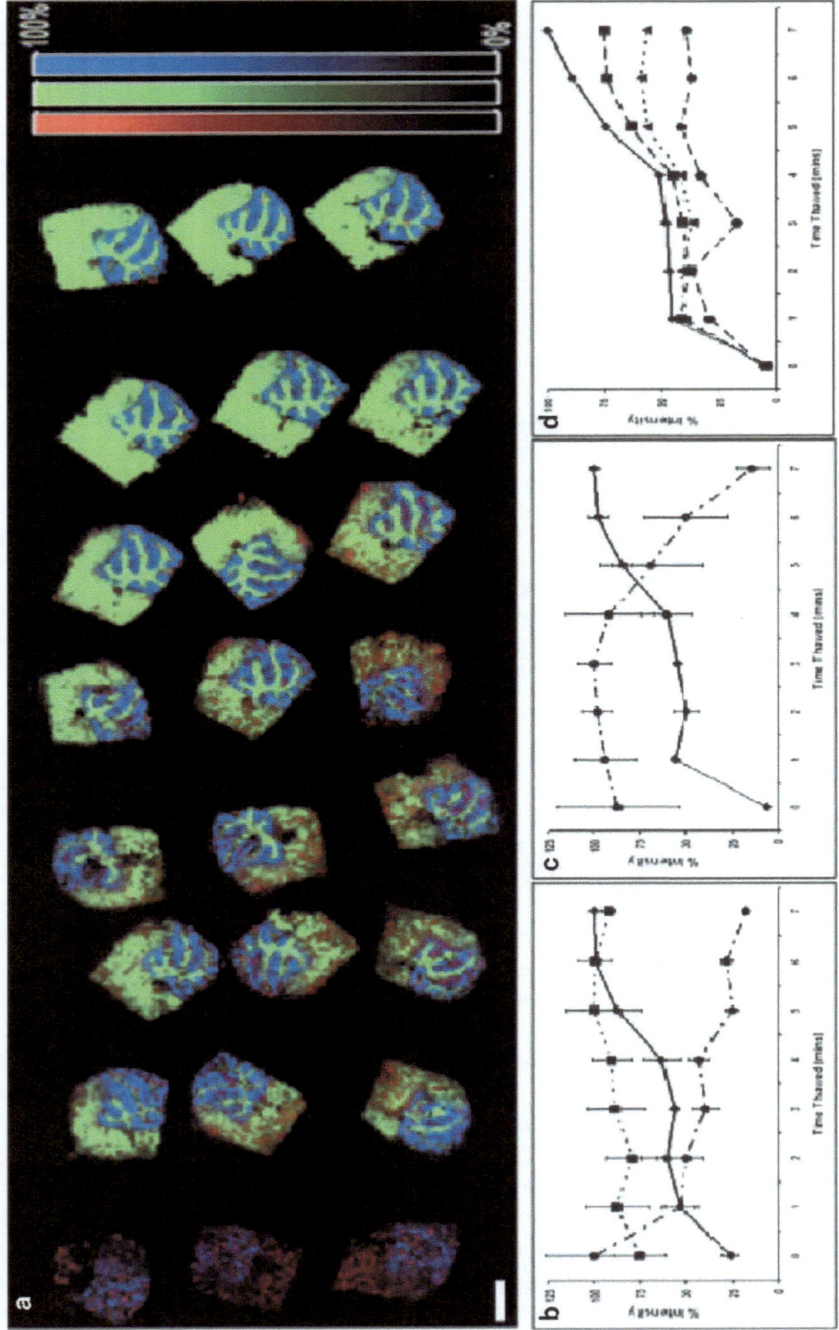

Fig. 4.6 MALDI-IMS images and plots of relative normalized intensities for a number of representative marker masses for 14-mm-thick mouse sections of the cerebellum thaw-mounted onto ITO-coated slides. Samples were manually spray-coated with 10 mg ml^{-1} cyano-4-hydroxy cinnamic

time course (*red*, decreasing; *blue*, no significant change; *cyan*, increasing) [19]. Therefore, tissue slices are preferably served at the next step – the washing step for protein measurement and the matrix application step for small molecule measurement – as rapidly as possible. Considerable care is required at these stages to facilitate comparisons among the biomarkers of different IMS experiments.

4.6 Rinsing of Tissue Sections with Organic Solvents

As mentioned in Chap. 2 of this book, the rinsing of tissue sections with organic solvents is required when detection targets include peptides and proteins. For the measurement of small molecules, this process is usually eliminated [23]. The rinsing step promotes the ionization of peptides and proteins, mainly by washing out phospholipids from the sections [1, 6]. It also plays a role in flushing out salts that could interfere with the crystallization of the matrix [1].

In earlier studies, several rinsing methods using ethanol solutions were described. The following are representative procedures:

- 70%:30% (v/v) ethanol/water (30-s immersion)×2 times (Fig. 4.7) [1]
- 70%:30% (v/v) ethanol/water (30-s immersion), 100% ethanol (15-s immersion) [24]
- 70%:30% (v/v) ethanol/water (30-s immersion), 100% ethanol (15-s immersion) [24]
- 90%:9%:1% (v/v/v) ethanol/water/glacial acetic acid (30-s immersion) [25]

On the other hand, Lemaire et al. examined washings with five different organic solvents – chloroform, xylene, toluene, hexane, and acetone – by dropping each of the solvents on a tissue sample (200 μl solvent per cm^2 of tissue), removing the solvent with a syringe, and then repeating the procedure. In each case, they found that the detection of peaks originating from proteins increased, compared to

---

Fig. 4.6 (continued) acid (CHCA) matrix. The figure represents three replicate experiments and a time course of 0, 1, 2, 3, 4, 5, 6, and 7 min at room temperature. Error bars on the graphs are SDs for $n=3$. **a** MALDI-IMS image of the replicate experiments vertically and time course horizontally. Intensity plots for three markers (*red*, decreasing; *blue*, no significant change; *cyan*, increasing) are shown, demonstrating typical changes observed over the time course. *Bar* 2 mm. **b** Graph of relative normalized intensity for three markers in the total tissue area showing a decrease (1007.8 6 0.5 Th, *filled circles*), increase (2407.3 6 0.5 Th, *filled diamonds*), or no significant change (1315.9 6 0.5 Th, *filled squares*). **c** Graph of relative normalized intensity for two markers in the total tissue area, one showing a decrease with a significantly different overall profile to that in (**b**) (950.9 6 0.5 Th, *filled circles*), and the other an increase very similar in rate to that in (**b**) (1835.5 6 0.5 Th, *filled diamonds*). **d** The apparent relative change in a marker (1835.5 6 0.5 Th) averaged across the whole tissue (*filled squares*), the cerebellum (*filled triangles*), the adjacent brainstem and neocortical tissue (*filled diamonds*), and molecular layer of the cerebellum (*filled circles*), showing how significant differences in the rate of change in specific regions can be masked when analyzing the whole tissue [19]. (Reprinted from Goodwin et al., Proteomics 8:3801–3808.)

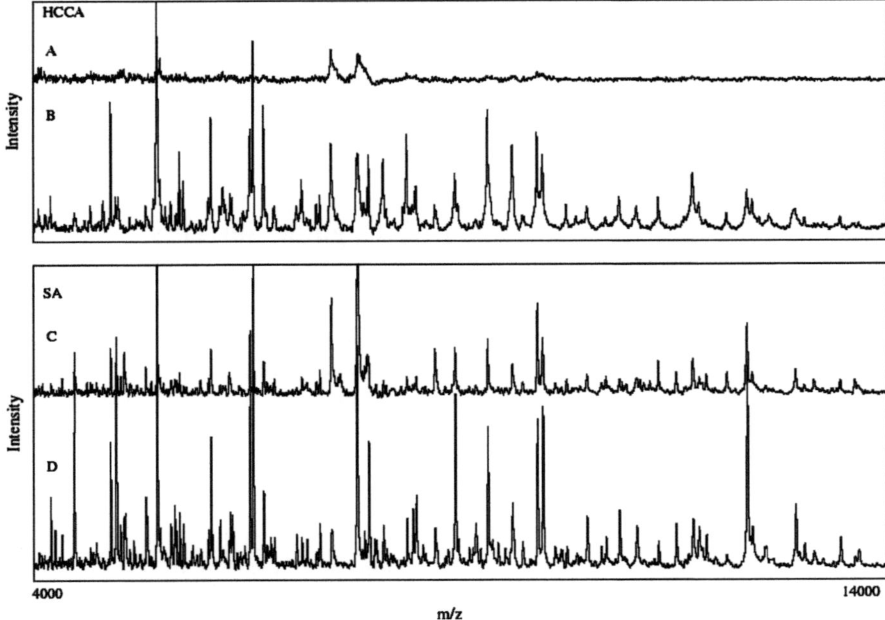

Fig. 4.7 Mass spectra obtained from mouse liver tissue before (**a, c**) and after (**b, d**) the washing process was completed [70%:30%, v/v, ethanol/water (30-s immersion)×2]. Cyano-4-hydroxycinnamic acid (CHCA) and sinapic acid (SA) were used as the matrix in the *upper* and the *lower panels*, respectively. (Reprinted from Schwartz et al., J Mass Spectrom 38:699–708.)

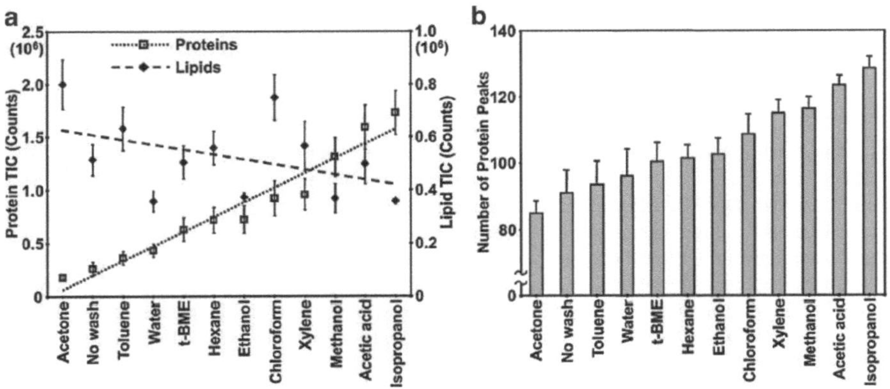

Fig. 4.8 (**a**) Total ion count (TIC) variations recorded from the MALDI-TOF MS protein profiles acquired from serial mouse liver tissue sections not washed or washed with different solvent systems in the *m/z* range 500–1,100 (lipid component) and the *m/z* range 2,000–25,000 (protein component). (**b**) Number of peak variations as a function of the same washes for the protein component [7] (Reprinted from Seeley et al., J Am Soc Mass Spectrom 19:1069–1077.)

untreated samples [6]. More recently, Seeley et al. demonstrated that the rinsing step enhances protein detection in terms of both the number of observed peaks and ion counts, and that this trend was inversely correlated with lipid detection (Fig. 4.8) [7]; after examining 12 different rinsing solvents, these authors established the following conditions as the most effective for protein analysis, with respect to the S/N ratio of protein signals, matrix crystal forms, and histological integrity of the tissues:

70%:30% (v/v) isopropanol/water (30-s immersion), followed by 90%:10% (v/v) isopropanol/water (30-s immersion) [7]

4.7 Staining of Tissue Sections Before IMS Measurement

As mentioned earlier, the use of ITO-coated materials makes microscopic observation possible. Invaluable expertise is acquired by examining the ion distribution images of IMS in conjunction with light microscopic images; however, simply viewing a frozen section under a microscope would merely reveal the indistinct outlines of the tissue's internal structural organization. When there are certain particular target areas or specific lesion sites to be analyzed by an experimenter, it is preferable that the analyzed sites are, initially, selectively determined by using appropriate dye methods. Chaurand et al. examined the mass spectra of tissue sections stained with certain different staining dyes and introduced staining methods that can be used in concert with IMS (Fig. 4.9) [5].

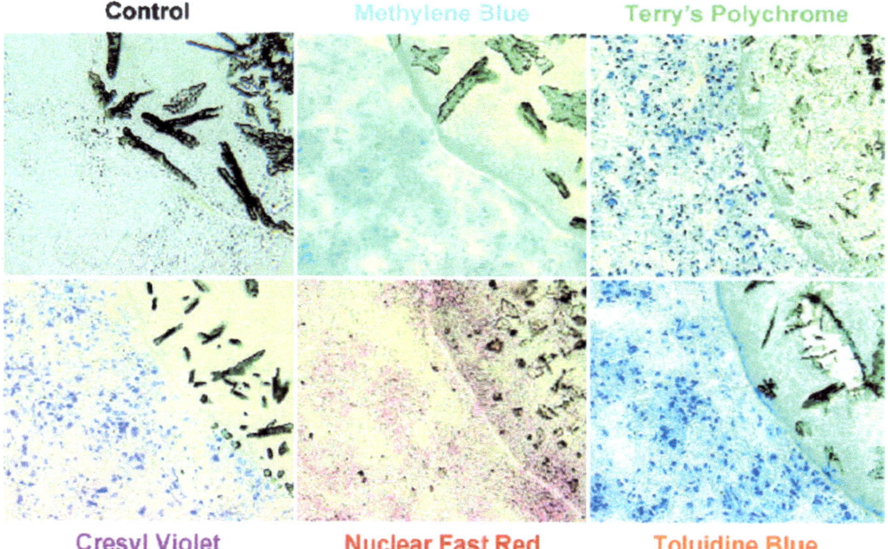

Fig. 4.9 Microscopic image (×400) of human glioma tissue sections that were stained with different dyes and onto which matrix was subsequently deposited. (From [3]) (Reprinted from Chaurand et al., Anal Chem 76(4):1145–1155.)

4.8   A Standard Procedure for Preparing Frozen Sections

4.8.1   Materials

The following items are needed when preparing frozen tissue sections:
- Cryostat
- Blades for cryostat
- OTC compound
- ITO glass slides

4.8.2   Methods

To prepare frozen tissue sections, execute the following procedure:
1. Place a glass slide inside the cryostat
2. Wipe off any oil remaining on the blade with ethanol and attach the blade to the cryostat
3. Pre-cool the forceps and brushes by placing them inside the cryostat
4. After the conditions for thin sectioning have been determined, i.e., the position of the anti-roll, temperature, and thickness (Note 1), perform thin sectioning of the sample
5. Keep the slice under the anti-roll for a while, and hold the end of slice with forceps or brushes when it no longer curls
6. Hold up the glass slide in the cryostat, and immediately place the slice on it
7. Keep performing thin sectioning and pasting
8. Cool a container filled with nitrogen gas in the cryostat, and immediately transfer the section to this container, and then store at $-80°C$ (Note 3)
9. When the sections are to be analyzed, take them out of the container, and then dry them in a stream of nitrogen gas (Note 4)
10. Serve the slice to the next process (washing process for protein analysis or matrix application for small molecule analysis)

4.8.3   Notes

1. Sections tend to curl up when the temperature is too high or if the anti-roll is placed inappropriately.
2. For the analysis of small molecules that might be oxidized, it is recommended to serve the section to the matrix application step immediately after it has been excised.
3. Insufficient drying causes the sample to peel off.

4.9 Conclusion

In this chapter, we briefly introduced sample preparation methodologies, particularly those used in the preparation of tissue sections. Because there are a number of key points in this process that can determine the success of an IMS experiment, we identified each of these and described the techniques used to handle them. However, there are still many aspects of these techniques to be investigated: for example, lipid-rich organs, including the brain, require a stringent washing protocol, whereas such a treatment might even result in the removal of proteins in other organs. We proposed a protocol that summarizes the aforementioned key points. The optimization of tissue preparation protocols is an ongoing initiative; thus, the inquisitive reader would be directed not only to the excellent work referenced in this chapter but also to the most recent literature.

References

1. Schwartz SA, Reyzer ML, Caprioli RM (2003) Direct tissue analysis using matrix-assisted laser desorption/ionization mass spectrometry: practical aspects of sample preparation. J Mass Spectrom 38:699–708
2. Chaurand P, Norris JL, Cornett DS, et al. (2006) New developments in profiling and imaging of proteins from tissue sections by MALDI mass spectrometry. J Proteome Res 5:2889–2900
3. Stoeckli M, Staab D, Schweitzer A (2006) Compound and metabolite distribution measured by MALDI mass spectrometric imaging in whole-body tissue sections. Int J Mass Spectrom 260(2–3):195–202
4. Sugiura Y, Shimma S, Setou M (2006) Thin sectioning improves the peak intensity and signal-to-noise ratio in direct tissue mass spectrometry. J Mass Spectrom Soc Jpn 54:4
5. Chaurand P, Schwartz SA, Billheimer D, et al. (2004) Integrating histology and imaging mass spectrometry. Anal Chem 76:1145–1155
6. Lemaire R, Wisztorski M, Desmons A, et al. (2006) MALDI-MS direct tissue analysis of proteins: improving signal sensitivity using organic treatments. Anal Chem 78:7145–7153
7. Seeley EH, Oppenheimer SR, Mi D, et al. (2008) Enhancement of protein sensitivity for MALDI imaging mass spectrometry after chemical treatment of tissue sections. J Am Soc Mass Spectrom 19:1069–1077
8. Altelaar AF, Klinkert I, Jalink K, et al. (2006) Gold-enhanced biomolecular surface imaging of cells and tissue by SIMS and MALDI mass spectrometry. Anal Chem 78:734–742
9. Garrett TJ, Prieto-Conaway MC, Kovtoun V, et al. (2006) Imaging of small molecules in tissue sections with a new intermediate-pressure MALDI linear ion trap mass spectrometer. Int J Mass Spectrom 260:11
10. Scherl A, Zimmermann-Ivol CG, Di Dio J, et al. (2005) Gold coating of non-conductive membranes before matrix-assisted laser desorption/ionization tandem mass spectrometric analysis prevents charging effect. Rapid Commun Mass Spectrom 19:605–610
11. Chaurand P, Schwartz SA, Caprioli RM (2002) Imaging mass spectrometry: a new tool to investigate the spatial organization of peptides and proteins in mammalian tissue sections. Curr Opin Chem Biol 6:676–681
12. Stoeckli M, Chaurand P, Hallahan DE, et al. (2001) Imaging mass spectrometry: a new technology for the analysis of protein expression in mammalian tissues. Nat Med 7:493–496
13. Goodwin RJ, Pennington SR, Pitt AR (2008) Protein and peptides in pictures: imaging with MALDI mass spectrometry. Proteomics 8:3785–3800

14. Andersson M, Groseclose MR, Deutch AY, et al. (2008) Imaging mass spectrometry of proteins and peptides: 3D volume reconstruction. Nat Methods 5:101–108
15. Shimma S, Sugiura Y, Hayasaka T, et al. (2008) Mass imaging and identification of biomolecules with MALDI-QIT-TOF-based system. Anal Chem 80:878–885
16. Hayasaka T, Goto-Inoue N, Sugiura Y, et al. (2008) Matrix-assisted laser desorption/ionization quadrupole ion trap time-of-flight (MALDI-QIT-TOF)-based imaging mass spectrometry reveals a layered distribution of phospholipid molecular species in the mouse retina. Rapid Commun Mass Spectrom 22:3415–3426
17. Routtenberg A, Tarrant S (1974) Synaptic morphology and cytoplasmic densities: rapid postmortem effects. Tissue Cell 6:777–788
18. Moleman P, Bruinvels J, Westerink BH (1977) Rapid post mortem changes in 3,4-dihydroxyphenylacetic acid (DOPAC), a dopamine metabolite, in rat striatum. J Neurochem 29:175–177
19. Goodwin RJ, Dungworth JC, Cobb SR, et al. (2008) Time-dependent evolution of tissue markers by MALDI-MS imaging. Proteomics 8:3801–3808
20. Rehncrona S, Westerberg E, Akesson B, et al. (1982) Brain cortical fatty acids and phospholipids during and following complete and severe incomplete ischemia. J Neurochem 38:84–93
21. Umemura A, Mabe H, Nagai H, et al. (1992) Action of phospholipases A2 and C on free fatty acid release during complete ischemia in rat neocortex. Effect of phospholipase C inhibitor and N-methyl-D-aspartate antagonist. J Neurosurg 76:648–651
22. Stavinoha WB, Weintraub ST, Modak AT (1973) The use of microwave heating to inactivate cholinesterase in the rat brain prior to analysis for acetylcholine. J Neurochem 20:361–371
23. Khatib-Shahidi S, Andersson M, Herman JL, et al. (2006) Direct molecular analysis of whole-body animal tissue sections by imaging MALDI mass spectrometry. Anal Chem 78:6448–6456
24. Aerni HR, Cornett DS, Caprioli RM (2006) Automated acoustic matrix deposition for MALDI sample preparation. Anal Chem 78:827–834
25. Groseclose MR, Andersson M, Hardesty WM, et al. (2007) Identification of proteins directly from tissue: in situ tryptic digestions coupled with imaging mass spectrometry. J Mass Spectrom 42:254–262

Chapter 5
Matrix Choice

Yuki Sugiura, Mitsutoshi Setou, and Daisuke Horigome

Abstract In this section, the choices of matrix compound and solvent composition appropriate for tissue IMS are reviewed. As is well known, it is very important to choose an appropriate matrix for successful imaging measurement. A practical choice of matrix depends upon the type of analyte involved. Until today, in traditional MALDI-MS, a large variety of compounds has been empirically tested for their suitability in playing the role of a matrix; today, researchers can choose from a relatively small number of established "organic chemical matrices" such as sinapic acid (SA), α-cyano-4-hydroxy-cinnamic acid (CHCA), and 2,5-dihydroxybenzoic acid (DHB), and they have proven to be useful matrices for MALDI-imaging measurement. On the other hand, in MALDI-IMS, it is still necessary to develop a new matrix because of the extremely complex chemistry on the tissue surface. We also introduce some novel organic matrices and the further use of nanoparticles as an alternative to organic matrices from recent literature.

5.1 Principle of Molecular Ionization

Two ionization methods – matrix-assisted laser desorption/ionization (MALDI) and secondary ion mass spectrometry (SIMS) – are widely used for performing IMS with tissue sections [1]. MALDI-MS can measure large mass ranges of ions in a tissue section, and it can also perform molecular identification via tandem mass

Y. Sugiura
Department of Bioscience and Biotechnology, Tokyo Institute of Technology,
4259 Nagatsuta-cho, Midori-ku, Yokohama, Kanagawa 226-8501, Japan

Y. Sugiura and M. Setou (✉)
Department of Molecular Anatomy, Hamamatsu University School of Medicine,
1-20-1 Handayama, Higashi-ku, Hamamatsu, Shizuoka 431-3192, Japan
e-mail: setou@hama-med.ac.jp

D. Horigome
Mitsubishi Kagaku Institute of Life Sciences, 11 Minamiooya, Machida, Tokyo, Japan

M. Setou (ed.), *Imaging Mass Spectrometry: Protocols for Mass Microscopy*,
DOI 10.1007/978-4-431-09425-8_5, © Springer 2010

spectrometry (MS^n). If compared, SIMS can ionize the $m/z < 1000$ range of ions; at that point, it can hardly perform molecular identification by tandem mass spectrometry. On the other hand, SIMS-based IMS has a much higher spatial resolution (a few hundred nanometers) than that of MALDI-IMS (dozens of micrometers), as a result of the tightly focused primary ion beam that is narrower than the UV-pulsed laser beams of MALDI (see details of SIMS in Part VIII).

MALDI is a soft ionization technique allowing the analysis of large biomolecules (biopolymers such as proteins, peptides, and sugars) and large organic molecules (such as polymers, dendrimers, and other macromolecules) [2]. It is carried out by co-crystallizing the analyte and matrix and ionization triggered by irradiating laser to the co-crystal. The matrix compound is excited by absorbing laser energy, which is converted into heat; the heat then evaporates part of the analyte molecules [3]. The matrix is then thought to transfer part of its charge to the analyte molecules, thus ionizing them while still protecting them from the disruptive energy of the laser.

Protonated and deprotonated molecules are generally designated as $[M+H]^+$ and $[M-H]^-$, respectively. If alkali metal ions such as sodium and potassium are contained in the co-crystal (as is often true for biological tissue samples), sodium adduct $[M+Na]^+$ and potassium adduct $[M+K]^+$ ions are also generated. Such a soft ionization technique was first reported by Tanaka et al. (1988), who enhanced the subsequent development of MS for biomolecules and large organic molecules [4]. Subsequent studies have excellently improved the soft ionization method, which had previously limited the molecular weight of the analyte to the extent of 10 kDa for peptides/proteins; those studies have developed the chemical matrices used today, which push the measurable mass range to around 100 kDa [2].

5.2 Choice of Matrices

The essential functions required for the matrix in measuring biological macromolecules in MS are as follows:

1. Isolate analyte molecules by dilution and prevent analyte aggregation
2. Stabilize the matrix–analyte co-crystal in the vacuum chamber
3. Absorb laser energy via electronic excitation
4. Disintegrate the condensed phase of the co-crystal without excessive destructive heating of the embedded analyte molecules

It is very important to choose an appropriate matrix for successful imaging. A practical choice depends upon the type of analyte involved. Up to now, for traditional MALDI-MS, a large variety of compounds has been empirically tested for their suitability in playing the role of a matrix; today, researchers can choose from a relatively small number of established "organic chemical matrices," e.g., benzoic or cinnamic acid derivatives, and these are also available for MALDI imaging measurement. For example, sinapic acid is commonly used for imaging of relatively high molecular weight proteins, whereas 2,5-dihydroxybenzoic acid (DHB) is applied to small organic compounds, such as lipids. The properties of the three major matrices used for the MALDI-MS analysis of tissue sections are summarized in Table 5.1.

On the other hand, in MALDI-IMS, it is constantly necessary to search for potential matrix compounds because of the extremely complex chemistry on the tissue surface. Improvements to the current chemical matrices in terms of mass resolution, ionization efficiency, and measurable molecular weight range are essential for development of IMS methodology, because direct tissue analyses generally lead to a lowered spectral quality – likely the result of the nature of the complex chemistry in direct tissue MALDI-MS, involving numerous factors (e.g., thickness, freezing date, or type of tissue). Below, we introduce some of these challenges.

5.2.1   Ionic Matrices

Ionic matrices comprising organic acids and organic bases, in particular, have attracted attention over the years [5]. By applying simple synthetic processes vis-à-vis acid–base reactions, solid ionic matrices can be produced. For example, by

Table 5.1   Three major matrices used with MALDI-MS

Matrix	SA	CHCA	DHB
Other name	• Sinapinic acid • 3,5-Dimethoxy-4-hydroxycinnamic acid	• α-Cyano-4-hydroxycinnamic acid	• 2,5-Dihydroxy benzoic acid
Structural formula			
MW	224.21	189.17	154.12
Chemical formula	$C_{11}H_{12}O_5$	$C_{10}H_7NO_3$	$C_7H_6O_4$
Solubility	• Low solubility in H_2O • Soluble in methanol/H_2O and polar organic solvents	• Low solubility in H_2O • Soluble in methanol/H_2O and polar organic solvents	• Soluble in H_2O • Soluble in methanol/H_2O and polar organic solvents
Feature	High signal-to-noise ratio		The quality of a mass spectrum largely depends on the quality of the matrix's crystal
Subject	Protein (4–30 kDa)	Lipid and peptides (~8 kDa)	Lipid and peptides (~5 kDa)

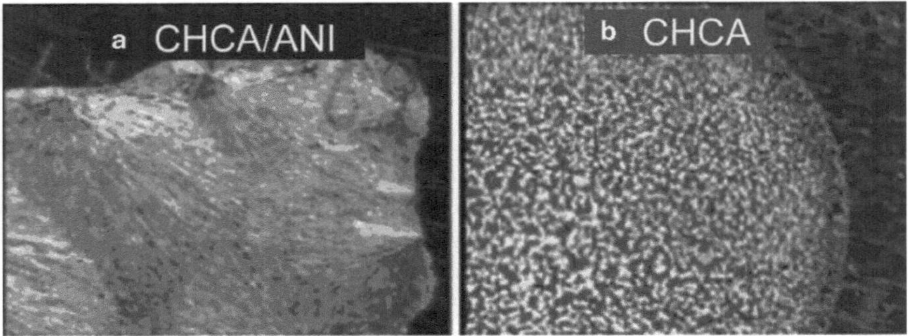

Fig. 5.1 Photographs of the matrix crystals of CHCA/ANI (**a**) in comparison with conventional CHCA (**b**). (Reprinted from Lemaire et al., Anal Chem 78(3):809–819.)

adding an equimolar amount of aniline (ANI) to the conventional matrix compound CHCA in methanol and subsequently evaporating the solvent, one can produce powdered CHCA/ANI. The solid ionic matrix is dissolved in the solvent [2:1 acetonitrile/0.01% trifluoroacetic acid (TFA)] at a concentration of 10 mg ml^{-1}. This ionic matrix can form extremely dense matrix crystals on a tissue section (Fig. 5.1) [6]. Compared to the conventional CHCA matrix, the CHCA/ANI matrix can achieve a much higher quality of IMS consequent to its better signal-to-noise (S/N) ratio (Fig. 5.2) [6]. Spectrum improvements attributed by the CHCA/ANI matrix resulted in better ion image quality than when using a conventional matrix (Fig. 5.3), in terms of increased signal detection and improved dynamic range of ion intensity, with reproducibility [6].

5.2.2 Challenges for Imaging of Primary Metabolites in m/z < 1,000 Region

Regarding imaging of low molecular weight compounds, one of the disadvantages of organic matrices is the number of mass peaks in the low m/z range. The low m/z region ($<m/z$ 1,000) of a MALDI spectrum contains a large population of ions from endogenous metabolites as well as matrix-related adduct clusters and fragments, which are clearly seen in the MALDI ion mobility spectrum obtained on the tissue section (Fig. 5.4) [7, 8]. Such high density of ions increases the risk for sharing the same mass window by matrix ions and analyte molecules.

Recently, 9-aminoacridine (9-AA) was reported to exhibit very few matrix interferences in the low-mass range ($m/z<500$) [10] and thus enables us to image primary metabolites in a MALDI imaging experiment [11, 12]. Benabdellah et al. reported that with appropriate sample preparation protocol, 9-AA exhibits almost no matrix interference, and they successfully detected and identified 13 primary metabolites (AMP, ADP, ATP, UDP-GlcNAc, etc.) on the rat brain section, in negative ion detection mode

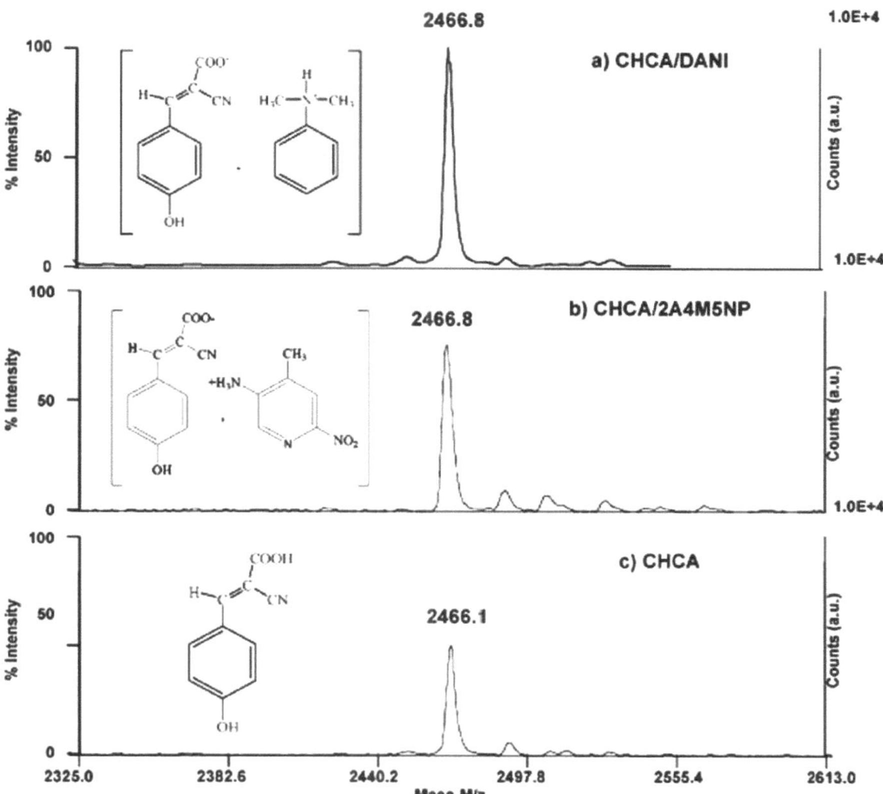

Fig. 5.2 Typical MALDI mass spectrum obtained for ACTH 18-39 (1 pmol) in the linear negative ion mode using CHCA/DANI (**a**), CHCA/2A4M5NP (**b**), and CHCA (**c**) as matrix. Ionic matrix formula is enclosed to each mass spectrum. (Reprinted from Lemaire et al., Anal Chem 78(3): 809–819.)

(Fig. 5.5) [12]. In addition, Burrell et al. also demonstrated that, by use of the 9-AA in positive ion detection mode, localization of sugar and phosphorylated metabolites such as glucose-6-phosphate can be clearly imaged in plant tissues (Fig. 5.6) [11].

These advances are quite important to develop MALDI-IMS as a practical tool for metabolite imaging in the clinical and biological field, because, until today, we did not have established imaging technique for such primary metabolites.

5.2.3   Nanoparticle-Based IMS

One of the critical limitations of the spatial resolution of MALDI-IMS is the size of the organic matrix crystal and the analyte migration during the matrix application process. To overcome these problems, our research group reported a nanoparticle-assisted

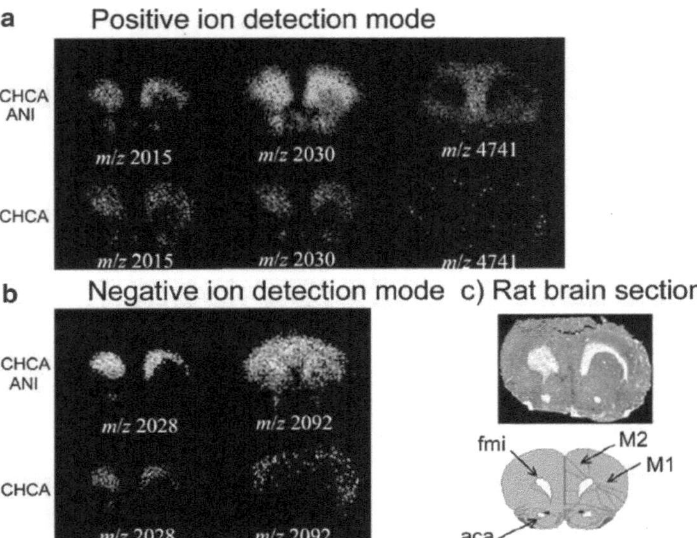

Fig. 5.3 MALDI-IMS using MALDI LIFT-TOF in reflector mode at 50-Hz repetition rate with ionic matrixes CHCA/ANI and CHCA in positive (**a**) and negative modes (**b**). MALDI imaging can be compared with rat brain anatomy. For CHCA/ANI and CHCA, acquisitions in both polarities were performed on the same rat brain cut (**c**) Optical image and corresponding atlas of rat brain section. (Reprinted from Lemaire et al., Anal Chem 78(3):809–819.)

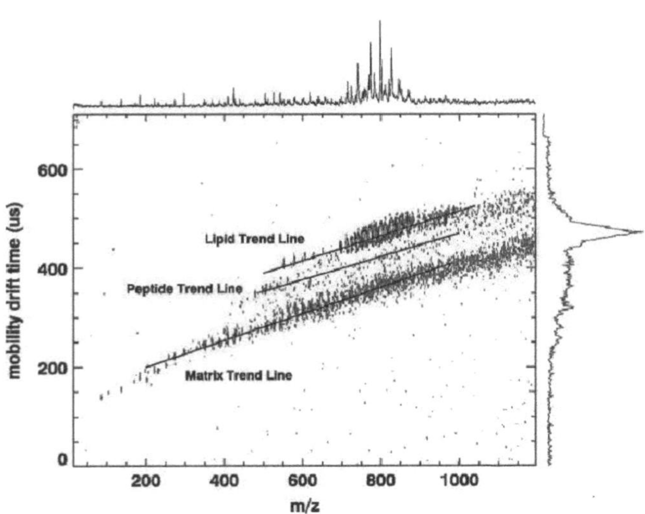

Fig. 5.4 MALDI ion mobility two-dimensional (2D) plot of a rat brain tissue section with DHB matrix in positive ion detection mode. Many of the peaks in the trend line identified as "matrix" can be assigned to DHB clusters or DHB clusters + potassium. (See details of MALDI ion-mobility MS in Chap. 17). (Reprinted from Jackson et al., J Mass Spectrom 42:1093–1098.)

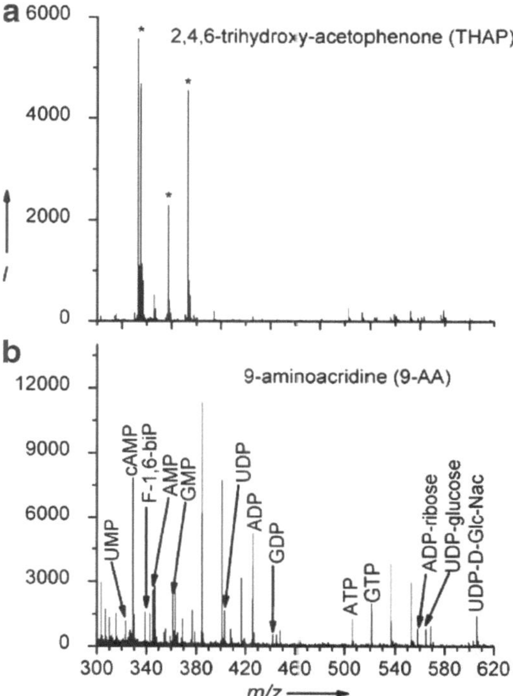

Fig. 5.5 MALDI-MS spectra in the negative ion mode acquired from a rat brain tissue section with matrix solution of (**a**) 2,4,6-trihydroxyacetophenone (THAP) and (**b**) 9-aminoacridine (10 mg ml^{-1}, in methanol) deposited on the section. *Asterisk*, matrix peaks. (Reprinted from Benabdellah et al., Anal Chem 81(13):5557–5560.)

laser desorption/ionization (nano-PALDI)-based IMS [13], in which the matrix crystallization process is eliminated [14]. In nano-PALDI, spatial resolution is not restricted by the crystal size but only by the instrumental factor (such as laser spot diameter); thus, the use of functionalized nanoparticle (fNP, $d = \sim 3.5$ nm) as matrix has enabled researchers to image compounds with high spatial resolution at the cellular level.

Figure 5.7 shows an overview of nano-PALDI. As already mentioned, SIMS-based IMS is useful for direct biomolecular analysis at high spatial resolution without interference from the matrix background signal; however, the typical SIMS technique has rarely been used for MS^n analysis [15] and is limited to low molecular analysis. In this regard, nano-PALDI enables researchers to ionize relatively heavy molecules even up to the insulin molecule (MW 5,773) (Figs. 5.7 and 5.8) [16]. As another important advantage, spraying fNP on the tissue surface did not alter the optical image of the tissue structure (Fig. 5.8g,h). Furthermore, its ability to eliminate the matrix-derived signals is important, for analysis of small molecules [14, 17].

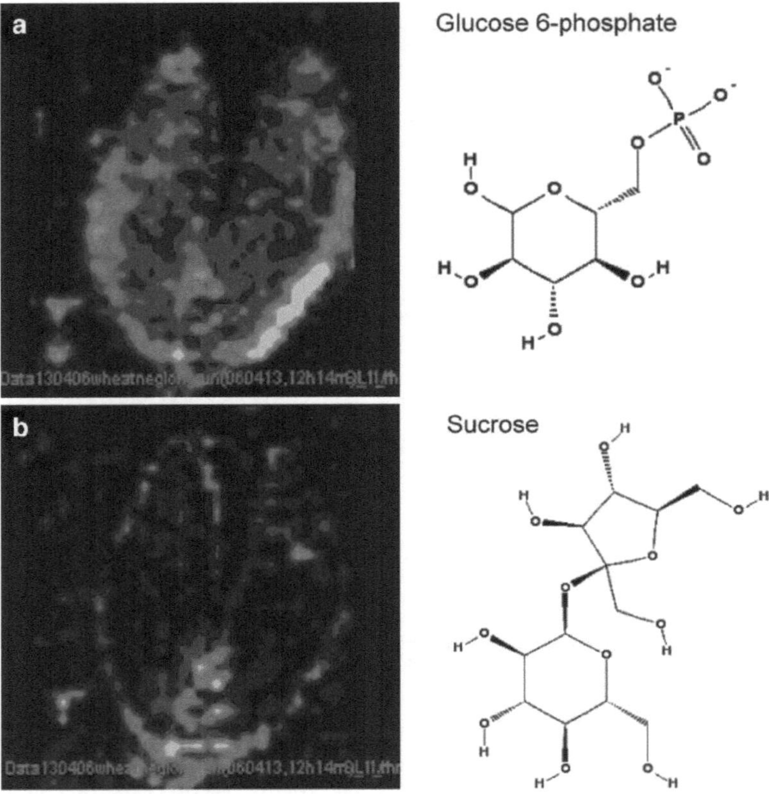

Fig. 5.6 IMS of glucose-6-phosphate and sucrose in developing wheat seeds: MALDI-IMS results for glucose 6-phosphate (**a**) and sucrose in wheat plant section (**b**). Intensity is plotted from *black* to *white* with white being the greatest signal. 9-AA (10 mg ml^{-1}, in acetone containing 0.1% TFA) was used as matrix. (Reprinted from Burrell et al., J Exp Bot 58:757–763.)

Because of their attractive features, nanoparticles are increasingly used as ionization-enhancing reagents as an alternative to organic matrices in IMS. Recently, gold nanoparticles ($d = \sim5.5$ nm) have been used in MS [18] and IMS [9]. Gold NPs ionize biomolecules that are difficult to detect using traditional organic matrices because of the unique ionization process [9].

5.3 Composition of Matrix Solvent

Composition of the matrix solvent and further matrix/solvent combination is also an important issue to optimize for successful imaging of the researcher's interest analyte. The goal of optimization the matrix solution is to effectively extract the

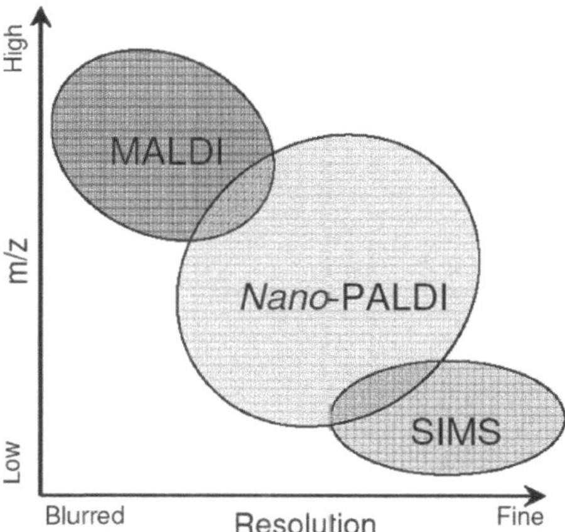

Fig. 5.7 Overview of nanoparticle-assisted laser desorption/ionization (nano-PALDI). Nano-PALDI can achieve high spatial resolution as in SIMS, and analysis of high molecules as in MALDI-IMS reprinted from [14]

analyte of interest while suppressing extraction of other molecular species from the tissues. Thus, optimal solvent composition will vary depending on the molecule to be analyzed as well as the type of tissue sample being analyzed [19].

Figure 5.9 demonstrates that organic solvent concentration in the matrix solution affects the detection of lipids and peptides. The results indicate that lipids and peptides could be efficiently extracted from tissue sections into organic and nonorganic solvents, respectively. Thus, for imaging of small molecules including peptides without a tissue-washing process (see Chap. 4), the results showed that a high composition of methanol was favorable for lipid detection whereas a low concentration solution was favorable for the detection of endogenous peptides. We additionally note that composition of the solvent also affects matrix–analyte co-crystal form; using DHB, needle-like (from which peptides were detected) changed into aggregates of smaller crystals (from which lipids were detected) (Fig. 5.9b). Because generation of a minute and homogeneous crystal layer among the tissue surface is required, when optimizing the matrix solvent, researchers should be aware of this issue as well as sensitivity of signal detection.

For protein analysis, one cannot also categorically describe which solvent is the best because the result of a solvent varies according to the type of tissue involved. For example, Shwartz et al. tested a series of saturated SA solutions in varying organic solvent/water combinations on a mouse liver section, and reported that an ethanol mixture is the best solvent so far for use with a mouse liver section whereas an acetonitrile mixture is the best one so far for use with a rat brain section [19]. Further, even for the same organ tissue sections, certain signal peaks have been observed only with an ethanol mixture solvent, whereas certain other signal peaks have been detected only when an acetonitrile mixture solvent was used (Fig. 5.10) [19]. Interestingly, those signal peaks could not be measured by using a three-in-one admixture (25:25:50 ethanol / acetonitrile / 0.1 % TFA in water).

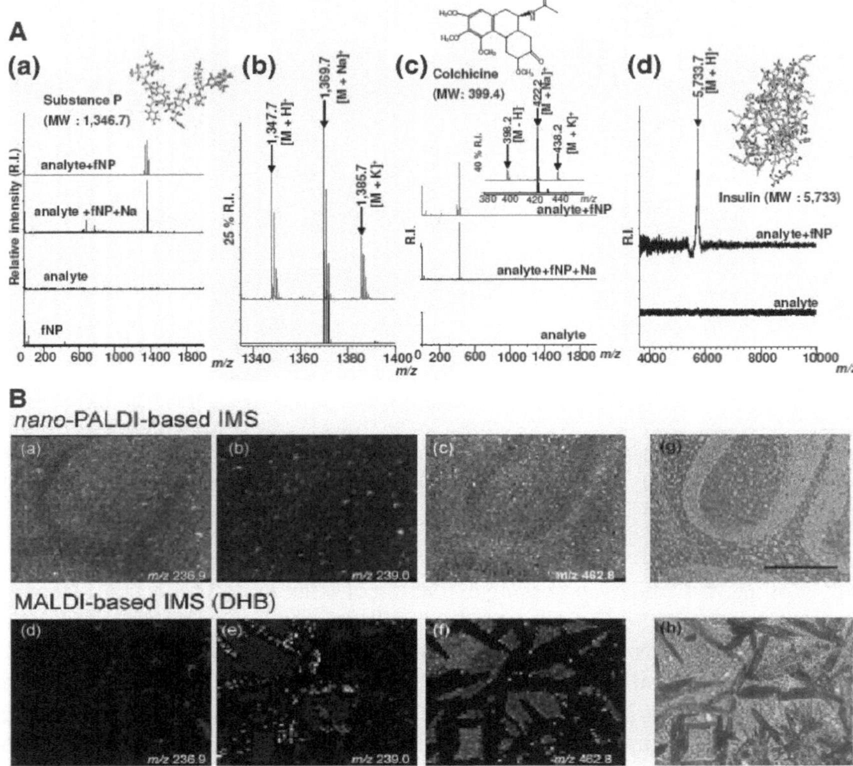

Fig. 5.8 Nano-PALDI MS and IMS. **A** The authors evaluated the usefulness of functional nanoparticles (NP) (fNP) in assisting ionization by using various analyte molecules. For peptide, in a spectra measured in an absence of external sodium acetate, signals for a proton, a sodium adduct, and a potassium adduct of substance P (MW 1,346.7) were observed. In the case of the colchicine drug (MW 399.4) (**c**) and insulin (MW 5,733) (**d**), fNP also facilitated ionization of the analytes, indicating that these can function as ionization-assisting reagents over a wide range of analytes. **B** Ion images obtained with the fNP (**a–c**) and with DHB (**d–f**) matrix. nano-PALDI can clearly provide each ion distribution images (**a–c**). However, from the DHB-sprayed section, several ions showed no significant distribution image (**e**), and were detected only from needle crystals (**e**) or the noncrystal region (**f**). Optical images of rat cerebellum tissues coated with the fNP (**g**) and with DHB (**h**) are also shown. *Bar* 500 μm (Reprinted from Taira et al., Anal Chem 80(12):4761–4766.)

As different solvent composition affects the overall protein profile, changing the concentration of TFA also changes the specific molecules measured. It has also been reported that a high concentration (>2%) of TFA can degrade a few signal peaks that could otherwise be detected with a solvent composition with a lower concentration of TFA [19], indicating that a high concentration of TFA is not recommended.

Overall, for protein analysis, a mixture of equal polar organic solvent (50% acetonitrile or ethanol) and 0.1–0.5% TFA in water is commonly used as the first choice of solvent and is recommended to be modified to obtain optimal performance.

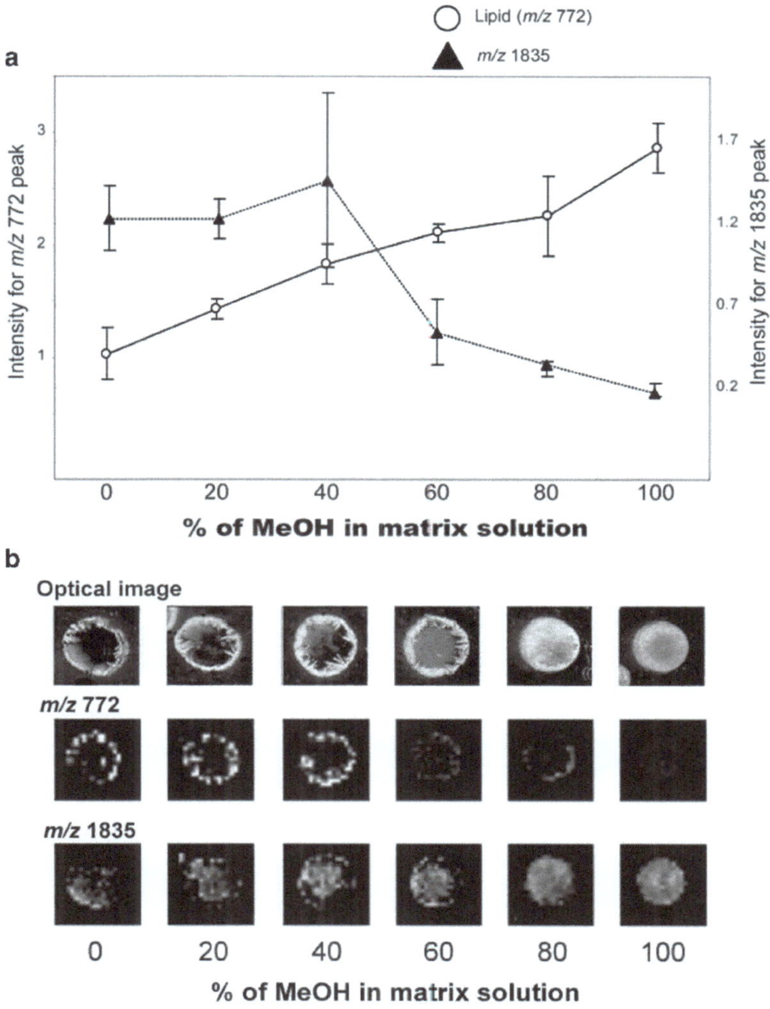

Fig. 5.9 Organic solvent concentration in the matrix solution affects the detection of lipids and peptides. **a** Matrix solutions with different water/methanol ratios (0%, 20%, 40%, 60%, 80%, and 100% methanol containing 0.1% TFA) were prepared. Then, 0.5 μl of each solution was spotted onto the brain homogenate sections ($n=3$). **b** Optical and ion images of the m/z 772 (lipid) and m/z 1,835, presumably derived from a peptide molecule. At high methanol concentrations, peptide detection was suppressed and lipid detection was favored (Reprinted from Sugiura et al., Rapid Commun Mass Spectrom 23:3269–3278.)

Finally, Table 5.2 summarizes the representative examples of matrix–solvent combination for IMS of various molecular species. Exploring the optimal matrix–solvent combination for each researcher's experiment, based on previous studies introduced here, is the key for successful IMS.

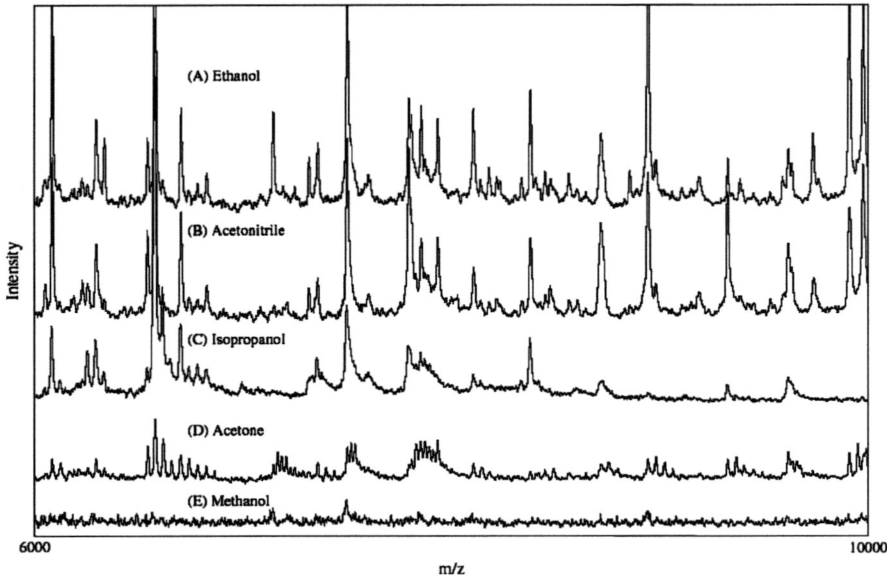

Fig. 5.10 Solvent composition affects the tissue protein profile. Saturated sinapinic acid matrix in solvents of 50:50 organic solvent/0.1% TFA in water as shown was deposited onto mouse liver tissue sections. The use of different solvent systems results in different protein profiles. Overall 50:50 ethanol/0.1% TFA resulted in the best protein profile for the mouse liver section, but 50:50 acetonitrile/0.1% TFA in water has been shown to give better profiles for other organs (such as rat brain). (Reprinted from Schwartz et al., J Mass Spectrom 38:699–708.)

Table 5.2 Representative examples of matrix/solvent combination for IMS of various molecular species

Analyte	Polarity	Matrix compound	Concentration (mg ml^{-1})	Solvent	Additives	Reference
Protein/peptide	POS	SA	Saturated	50% Acetonitrile, 0.1% TFA, 50% ethanol 0.1% TFA	–	Schwartz et al. 2003 [19]
Protein/peptide	POS	SA	25	50% Acetonitrile, 0.1% TFA	–	Seeley et al. 2008 [20]
Protein/peptide	POS	CHCA	10	50% Ethanol, 0.1% TFA	–	Altelaar et al. 2007 [21]
Drug	POS	CHCA	10	50% Acetonitrile, 0.1% TFA	–	Hopfgartner et al. 2009 [22]
Drug	POS	CHCA	25	Ethanol, 1% TFA	–	Atkinson et al. 2007 [23]
Phospholipids	POS	DHB	40	70% MeOH	Sodium acetate	Garrett et al. 2006 [7]
Phospholipids	POS/NEG	DHB	50	70% MeOH	Potassium acetate	Sugiura et al. 2009 [24]
Phospholipids	POS/NEG	MBT[a]	Saturated	MeOH	Cesium chloride	Astigarraga et al. 2008 [25]
Phospholipid (cardiolipin)	NEG	DHA[b]	30	50% Ethanol	Cesium iodide	Wang et al. 2007 [26]
Sugar/ phosphorylated metabolites	POS	9-AA	10	Acetone, 0.1% TFA	–	Burrell et al. 2007 [11]
nucleotides	NEG	9-AA	10	MeOH	–	Benabdellah et al. 2009 [12]

POS positive ion detection mode, *NEG* negative ion detection mode

[a]MBT, 2-mercaptobenzothiazole

[b]DHA, 2,6-dihydroxyacetophenone

References

1. McDonnell LA, Heeren RM (2007) Imaging mass spectrometry. Mass Spectrom Rev 26:606–643
2. Karas M, Bachmann D, Bahr U, et al. (1987) Matrix assisted ultraviolet-laser desorption of non-volatile compounds. Int J Mass Spectrom Ion Processes 78:53–68
3. Dreisewerd K (2003) The desorption process in MALDI. Chem Rev 103:395–426
4. Tanaka K, Waki H, Ido Y, et al. (1988) Protein and polymer analyses up to m/z 100,000 by laser ionization time-of flight mass spectrometry. Rapid Commun Mass Spectrom 2:151–153
5. Armstrong DW, Zhang LK, He L, et al. (2001) Ionic liquids as matrixes for matrix-assisted laser desorption/ionization mass spectrometry. Anal Chem 73:3679–3686
6. Lemaire R, Tabet JC, Ducoroy P, et al. (2006) Solid ionic matrixes for direct tissue analysis and MALDI imaging. Anal Chem 78:809–819
7. Garrett TJ, Prieto-Conaway MC, Kovtoun V, et al. (2006) Imaging of small molecules in tissue sections with a new intermediate-pressure MALDI linear ion trap mass spectrometer. Int J Mass Spectrom 260:11
8. Cornett DS, Frappier SL, Caprioli RM (2008) MALDI-FTICR imaging mass spectrometry of drugs and metabolites in tissue. Anal Chem 80:5648–5653
9. Jackson SN, Ugarov M, Egan T, et al. (2007) MALDI-ion mobility-TOFMS imaging of lipids in rat brain tissue. J Mass Spectrom 42:1093–1098
10. Amantonico A, Oh JY, Sobek J, et al. (2008) Mass spectrometric method for analyzing metabolites in yeast with single cell sensitivity. Angew Chem (Int Ed) 47:5382–5385
11. Burrell M, Earnshaw C, Clench M (2007) Imaging matrix assisted laser desorption ionization mass spectrometry: a technique to map plant metabolites within tissues at high spatial resolution. J Exp Bot 58:757–763
12. Benabdellah F, Touboul D, Brunelle A, et al. (2009) In situ primary metabolites localization on a rat brain section by chemical mass spectrometry imaging. Anal Chem 81, 5557–5560
13. Moritake S, Taira S, Sugiura Y, et al. (2009) Magnetic nanoparticle-based mass spectrometry for the detection of biomolecules in cultured cells. J Nanosci Nanotechnol 9:169–176
14. Taira S, Sugiura Y, Moritake S, et al. (2008) Nanoparticle-assisted laser desorption/ionization based mass imaging with cellular resolution. Anal Chem 80:4761–4766
15. McMahon JM, Short RT, McCandlish CA, et al. (1996) Identification and mapping of phosphocholine in animal tissue by static secondary ion mass spectrometry and tandem mass spectrometry. Rapid Commun Mass Spectrom 10:335–340
16. Moritake S, Taira S, Sugiura Y, et al. (2008) Magnetic nanoparticle-based mass spectrometry for the detection of biomolecules in cultured cells. J Nanosci Nanotechnol. 2009 (1):169–76.
17. Ageta H, Asai S, Sugiura Y, et al. (2008) Layer-specific sulfatide localization in rat hippocampus middle molecular layer is revealed by nanoparticle-assisted laser desorption/ionization imaging mass spectrometry. Med Mol Morphol 42:16–23 (2009)
18. McLean JA, Stumpo KA, Russell DH (2005) Size-selected (2–10 nm) gold nanoparticles for matrix assisted laser desorption ionization of peptides. J Am Chem Soc 127:5304–5305
19. Schwartz SA, Reyzer ML, Caprioli RM (2003) Direct tissue analysis using matrix-assisted laser desorption/ionization mass spectrometry: practical aspects of sample preparation. J Mass Spectrom 38:699–708
20. Seeley EH, Oppenheimer SR, Mi D, et al. (2008) Enhancement of protein sensitivity for MALDI imaging mass spectrometry after chemical treatment of tissue sections. J Am Soc Mass Spectrom 19:1069–1077
21. Altelaar AFM, Taban IM, McDonnell LA, et al. (2007) High-resolution MALDI imaging mass spectrometry allows localization of peptide distributions at cellular length scales in pituitary tissue sections. Int J Mass Spectrom 260:9
22. Hopfgartner G, Varesio E, Stoeckli M (2009) Matrix-assisted laser desorption/ionization mass spectrometric imaging of complete rat sections using a triple quadrupole linear ion trap. Rapid Commun Mass Spectrom 23:733–736

23. Atkinson SJ, Loadman PM, Sutton C, et al. (2007) Examination of the distribution of the bioreductive drug AQ4N and its active metabolite AQ4 in solid tumours by imaging matrix-assisted laser desorption/ionisation mass spectrometry. Rapid Commun Mass Spectrom 21:1271–1276

24. Sugiura Y, Konishi Y, Zaima N, et al. (2009) Visualization of the cell-selective distribution of PUFA-containing phosphatidylcholines in mouse brain by imaging mass spectrometry. J Lipid Res J Lipid Res 50: 1776–1788

25. Astigarraga E, Barreda-Gomez G, Lombardero L, et al. (2008) Profiling and imaging of lipids on brain and liver tissue by matrix-assisted laser desorption/ ionization mass spectrometry using 2-mercaptobenzothiazole as a matrix. Anal Chem 80:9105–9114

26. Wang HY, Jackson SN, Woods AS (2007) Direct MALDI-MS analysis of cardiolipin from rat organs sections. J Am Soc Mass Spectrom 18:567–577

27. Suguira Y, Setou M (2009) Selective imaging of positively charged polar and nonpolar lipids by optimizing matrix solution composition. Rapid Commun Mass Spectrom 23(20): 3269–3278

Chapter 6
Methods of Matrix Application

Yuki Sugiura, Mitsutoshi Setou, and Daisuke Horigome

Abstract The matrix application procedure is the most critical step of the sample preparation procedure, because it affects co-cystalization of the matrix-analyte complex formed on the tissue surface, thus the quality of the mass spectra and the subsequent imaging results. Several methods have been used for matrix application: (1) manually spraying the matrix solution with a nebulizer sprayer, (2) automatically depositing an array of small droplets of matrix solution with robotic devices, and (3) sublimating the matrices under reduced pressure and elevated temperature. In this chapter, we introduce an overview of these procedures and then continue with details of technical points, especially of methods (1) and (2). Lastly, a Further advanced application method, which is the spray-droplet method to increase signal sensitivity, is also described.

6.1 Introduction

For the newcomer to imaging mass spectrometry (IMS), matrix application may be the most difficult step of sample preparation because the quality of the mass spectra and the subsequent imaging results depend largely on this step. Today, several methods, in particular three major methods, have been employed for matrix application, as presented below (Fig. 6.1):

Y. Sugiura
Department of Bioscience and Biotechnology, Tokyo Institute of Technology,
4259 Nagatsuta-cho, Midori-ku, Yokohama, Kanagawa 226-8501, Japan

Y. Sugiura and M. Setou (✉)
Department of Molecular Anatomy, Hamamatsu University School of Medicine,
1-20-1 Handayama, Higashi-ku, Hamamatsu, Shizuoka 431-3192, Japan
e-mail: setou@hama-med.ac.jp

D. Horigome
Mitsubishi Kagaku Institute of Life Sciences, 11 Minamiooya, Machida, Tokyo, Japan

M. Setou (ed.), *Imaging Mass Spectrometry: Protocols for Mass Microscopy*,
DOI 10.1007/978-4-431-09425-8_6, © Springer 2010

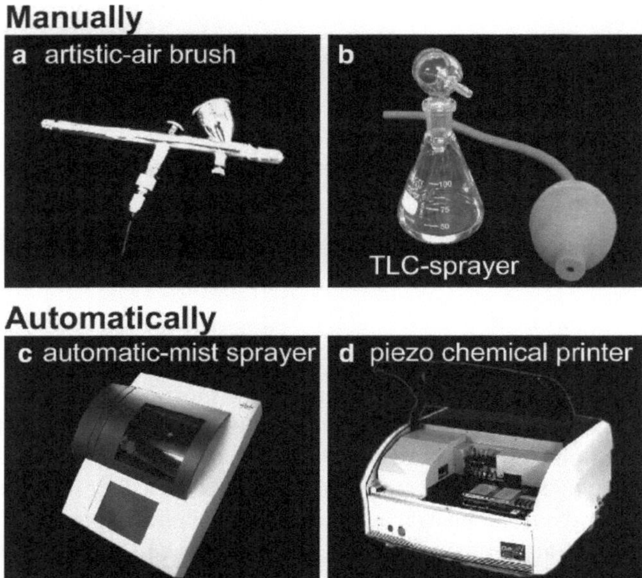

Fig. 6.1 Photograph of matrix application instruments. Current matrix application methodologies can be subdivided into manual operating methods (**a, b**) and automatic methods (**c, d**)

1. Manual spraying of the matrix solution with an artistic air-brush and thin layer chromatography (TLC) sprayer
2. Automatic depositing of an array of small matrix solution droplets with robotic devices [1, 2]
3. Sublimating the matrices under reduced pressure and elevated temperature [3, 4]

Development of other approaches to improve the homogeneity of applied crystal layer, reproducibility, and signal sensitivity is still ongoing [5], but here we introduce an overview and technical points of the three major methods, especially of methods (1) and (2), from our experience.

6.2 Manual Spray-Coating

The spray-coating method is one of the most frequently used methods. In this process, an entire tissue section can be homogeneously coated with relatively small crystals. For this operation, several instruments including TLC sprayers and artistic air-brushes are available (see Fig. 6.1a–c).

Although this method is familiar, with a convenient procedure, it requires skillful operation because of the numerous parameters of the hand-operation of an airbrush. Precise control of them is essential for reproducible IMS experiments (Fig. 6.2). If there is an excess of matrix solution on the tissue, an inhomogeneous

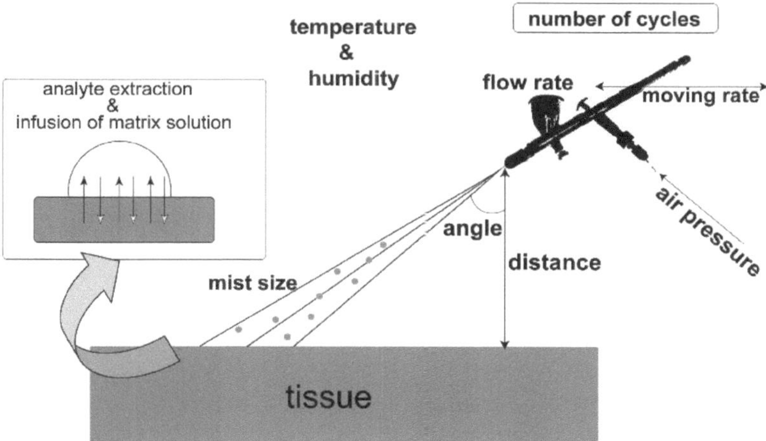

Fig. 6.2 Representative parameters of spraying operation with an air-brush that must be controlled among the trials for a reproducible IMS experiment

crystal with migrated analytes from its original location can be formed; on the other hand, if the amount of sprayed matrix is insufficient and the matrix solution evaporates without sufficiently moisturizing the tissue section, analytes cannot be extracted adequately from the tissue section. Moreover, the operation should be performed at constant room temperature and humudity for day-to-day reproducibility. Beginners are recommended to practice this spraying operation until homogeneous matrix spraying with reproducibility can be achieved.

Alternatively, one can use automatic spraying devices such as ImagePrep (Bruker Daltonics) (see Chap. 7) and a robotic sprayer (LEAP Technologies, Carrboro [6]), which achieve high reproducibility independent of the experimenter's proficiency.

6.2.1 Protocol of Spraying Operation with an Artistic Air-Brush

To achieve optimal spraying, we recommend to (1) minimize ejected droplet size, to accelerate the drying rate and generate minute crystal, (2) keep the air-brush in a position at a constant distance from the tissue, and (3) gradually move the air-brush horizontally. A key point to bear in mind when executing this method is to maintain equilibrium between two rates: the rate at which a fine aerosol of air-brushed matrix solution slightly moisturizes the tissue section, which facilitates analyte extraction, and another rate at which quick solvent evaporation which prevents analyte migration (see Fig. 6.2, inset).

Other important parameters to care for practical spraying include (1) the size of the droplet, (2) the amount of mist, (3) the angles and distances between the spray nozzle and the tissue section, and (4) laboratory temperature and humidity. These parameters are described below in detail.

6.2.2 Preparation of Matrix Solution

1. Weigh appropriate amount of matrix compound and put it into an organic solvent-tolerant microtube
2. Add 1.0 ml solvent with organic solvent-tolerant pipette tips
3. Thoroughly dissolve the matrix compound by vortexing or performing brief sonication
4. Store at room temperature until use

   Note:

- Microtubes with organic solvent-tolerant properties can be purchased from Eppendorf.
- Use HPLC- or LC/MS-grade solvent for preparing matrix solution.
- Preparing the matrix solution on the day of experiment is recommended.

6.2.3 Matrix Application with an Artistic Air-Brush

1. For test-spraying, pour approximately 1.0 ml solvent without matrix into the air-brush.
2. Optimize the size of the ejected droplet, the amount of mist, and the angles and distances between the nozzle and the sample.
3. Remove the solvent from the air-brush.
4. Mask the areas outside the tissue section which is mounted on the ITO glass slide, with a masking tape.
5. Fix the masked ITO glass slide onto a perpendicular board (Fig. 6.3).
6. Pour the prepared matrix solution into the air-brush and spray onto the tissue section on the ITO glass slide (Fig. 6.3).
7. As a rough guide for optimal coating, spray approximately 0.5–1.0 ml matrix solution for one ITO slide glass.
8. Remove the masking tape from the ITO glass slide after spraying, and place the sample in an airtight container with dry silica gel.
9. Perform the IMS measurement as soon as possible to minimize the progress of sample damage.
10. Figure 6.4 shows examples of good and unfavorable spraying results by spraying 2,5-dihydroxy benzoic acid (DHB) matrix solution (50 mg ml^{-1}, 70% MeOH containing 10 mM potassium acetate).

    Notes

- Choice of an air-brush
  In general, two properties are required for the air-brush: (1) adjustable droplet size and control of mist volume, and (2) it must be organic solvent tolerant to prevent the contamination of instrument derived component. We prefer a metal air-brush with a 0.2-mm nozzle, i.e., Procon Boy FWA Platinum (Mr. Hobby, Tokyo, Japan), because of its simple and easy-to-handle design.

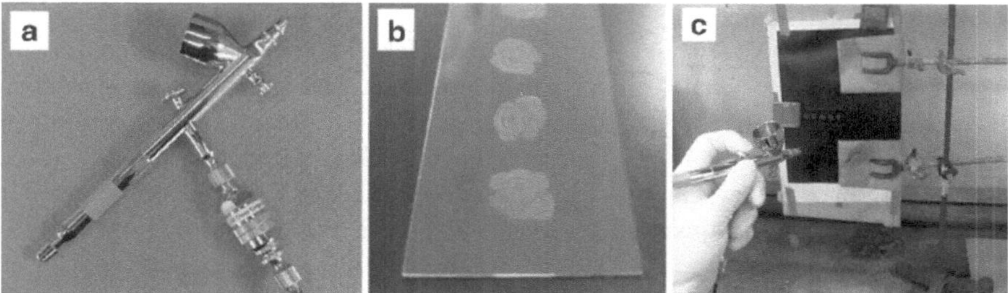

Fig. 6.3 Photographs of (**a**) an air-brush, (**b**) mouse cerebellum sections on an ITO-coated glass slide, and (**c**) spraying operation onto an ITO-coated glass slide fixed on a vertical board in the fume hood

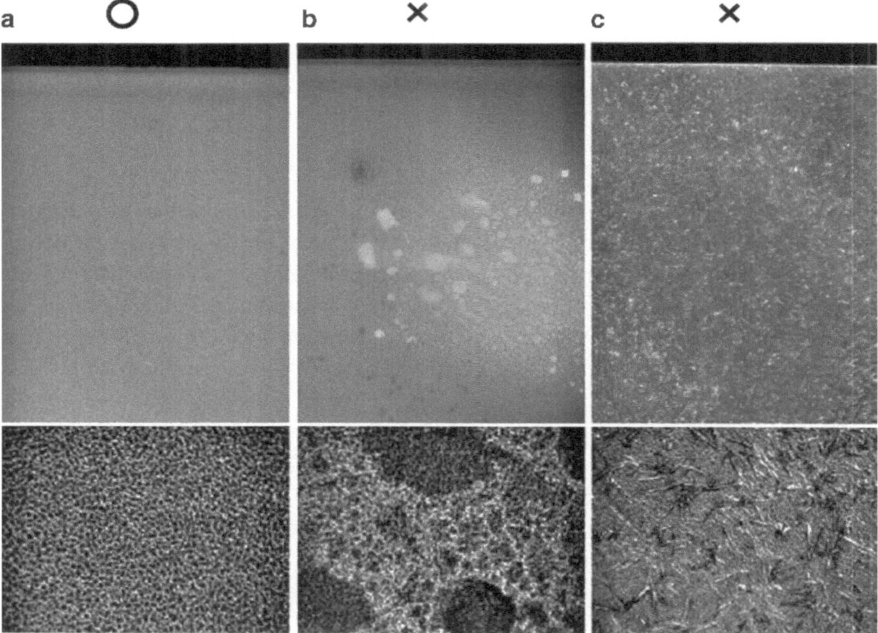

Fig. 6.4 Examples of results of DHB spray-coating with the air-brush. (**a**) A properly handled spray-coating step created a uniform matrix crystal layer, and awareness of certain technical points leads to a successful coating step. (**b**) Too small a distance between the air-brush and the tissues (<10 cm) often creates large droplets of matrix solution and results in inhomogeneous crystals. (**c**) Humidity is also an important factor. Room humidity was held under 25% at room temperature (25°C). High humidity tends to cause formation of needle-like crystals (>80%, at room temperature). The *upper panel* shows stereoscopic microscope images, and the *lower panel* shows phase-contrast microscopic images of the matrix layer formed on the glass slides. (From [7]) (Reprinted from Sugiura et al., J Lipid Res 50:1776–1788.)

- Working environment
  Air ventilation: Work under a hood, to avoid inhalation of matrix aerosols.
  Air conditioning: Control the room temperature and humidity, to reproduce homogeneous matrix crystallization. In general, we keep the air-conditioned room temperature at 23–25°C and at a low humidity of under 30%.
  Lighting: Matrix crystals that form on a tissue section look like white powder under bright light. Adjust lighting to easily observe crystal condition (i.e., homogeneity and color).
- Spray practices
  Optimize the parameters of the air-brush in advance by spraying the matrix solution onto a brown KimTowel. The purpose of this practice is to confirm the drying rate of the matrix solution at each experiment, which can be slightly affected even by the weather. As a rough guide, the optimal drying rate of a sprayed KimTowel is 1–2 s. A wet KimTowel becomes darker brown in color, thus indicating its state of dryness. In another way, the use of black paper is helpful because the white matrix crystals formed on the paper tell one how homogeneous a crystal layer was formed.

6.2.4 Other Notes

- IMS measurements should be performed as soon as possible after matrix application, regardless of coating method. We have observed apparent disturbance of the signal intensity of the mass spectra from the sample after a lapse of 48 h at room temperature. If we must store a sample, we place it in an airtight container with dry silica gel under nitrogen atmosphere after matrix application. Store it in a cold room (4°C) and use the sample within 24 h, preferably.

6.3 Automatic Dispensing of Small Droplets by the ChiP Instrument

Manual pipetting to deposit the matrix solution onto a tissue section is an effective technique to direct tissue profiling. It offers a higher ion yields than the spraying method because it allows better extraction of analyte from a tissue section [8]. However, in general, a dispensed matrix droplet makes a spot >1 mm in diameter on a tissue surface because of the lower limit of manual pipetting volume (500 nl) with an ordinary micropipette. Such large-sized spots are not sufficient in performing precise, high-resolution IMS; therefore, technical improvements achieving the dispensation of droplets as small as possible are required.

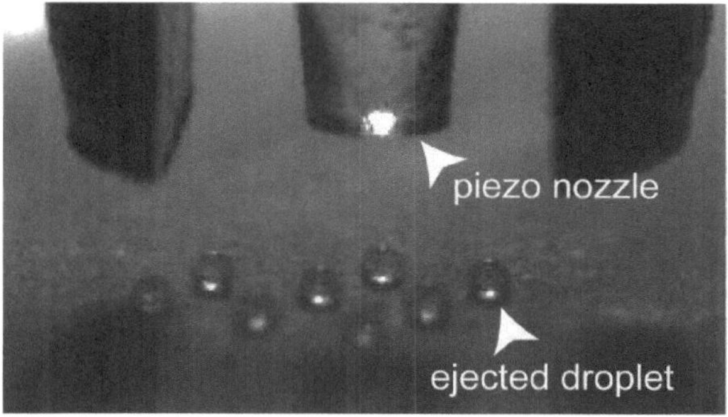

Fig. 6.5 Example representation of micro-ejection of small droplets by ChIP-1000 instrument

In response to these needs, the Chemical Inkjet Printer (ChIP-1000), an inkjet printer-like device equipped with a piezoelectric reagent dispensing system, has been developed by Shimadzu Corporation. The ChIP-1000 can dispense picoliter volumes of matrix solution onto a tissue section, using four printing heads and nozzles that have no contact with the surface of tissue sections. An example photograph of the micro-ejection is shown as Fig. 6.5. Using ChIP-1000, a spot with a diameter of 100–250 µm can be made on a tissue section by dispensing approximately 10 droplets of matrix solution of 87 pl with diameters of ~55 µm each. Therefore, it can create a two-dimensional (2D) matrix-spot array on the surface of tissue sections at intervals of 100 µm and several tens of micrometers. The adjacent spots are made independently, so that there is very little possibility of cross-contamination; thus, the distances between them define the image resolution of the IMS. On the other hand, image resolution with the spray-coating method depends upon the diameter of the laser beam of the mass spectrometer, which is at present approximately 10–100 µm. Thus, the droplet method with the automatic pipetting device can achieve an IMS image resolution that is in no way inferior to that obtained via the spray-coating method. The Acoustic Reagent Multispotter (ARM) is also known as an apparatus that has a unique nozzle-free ejector that dispenses picoliter volumes of reagent [1, 9].

6.3.1 On-Tissue Digestion and Matrix Application with the ChIP-1000 Instrument

Below, operation protocols and experimental examples using ChIP-1000 are provided. The benefit of methods that use an automatic minute-drop-dispensing device is that one can dispense drops with a high level of accuracy, vis-à-vis position and volume control. Also, one can dispense different reagent solutions at the same position in several batches. Taking advantage of this feature, one can perform the on-tissue digestion of proteins by undertaking a tandem approach, i.e., dispensing a protein-degrading enzyme solution and matrix solution in series, on the same spot. The on-tissue digestion protocol (see also Chap. 2) and experimental examples are described below.

6.3.1.1   Preparation of Trypsin Stock Solution

- A lyophilized powder of trypsin (Trypsin Gold, Mass Spec Grade) can be purchased from Promega Corporation.
- Reconstitute the powder in 50 mM acetic acid, to prepare 0.5 mg ml^{-1} trypsin stock solution; store at $-20°C$ for up to 1 month. For long-term storage, freeze reconstituted trypsin at $-70°C$.
- For use as a working solution, add 200 μl 100 mM NH_4HCO_3 to 40 μl trypsin stock solution, which results in a final concentration of 0.083 mg ml^{-1} ($pH < 8$).

6.3.1.2   Matrix Application Using the ChIP-1000

1. Start up the ChIP-1000 (Fig. 6.6a).
2. Dehydrate a rat brain section, e.g., 12 μm thick and mounted on an ITO slide glass [or a metal target plate for matrix-assisted laser desorption/ionization-mass spectrometry (MALDI-MS)].
3. Immerse the section in 70% ethanol for 30 s, and dry to remove lipids and salts. Repeat this operation twice.
4. Immerse the section in 90% ethanol/9% glacial acetic acid/1% water for 30 s, to fix the proteins in the tissue section.
5. Dehydrate the section in a vacuum desiccator.
6. Place the section on the sample stage of the ChIP-1000 (Fig. 6.6b).
7. Pour reagent solution into the print head (Fig. 6.6c).
8. Scan an optical image of the section with built-in scanner.
9. Create a file that designates printing positions.

When tryptic-digested proteins are measured using on-tissue digestion, the following six steps are also taken:

10. Dispense a 100-pl volume of 0.083 mg ml^{-1} trypsin solution and deposit five drops (0.5 nl) in one spot.
11. Dispense at each position period.
12. Iterate steps (10) and (11) 30 times at intervals of approximately 8 min, while the previously deposited spot dries. Achieve a total drop volume of 15 nl at each position.
13. The proteolysis reaction progresses during step (12) (approximately 4 h in total).
14. Dispense the matrix solution of 25 mg ml^{-1} DHB in 50% methanol/49.5% water/0.5% trifluoroacetic acid (TFA) in the same way as in steps (10) and (11). Iterate 20 times, to achieve a total drop volume of 10 nl at each position.
15. Perform IMS measurement.

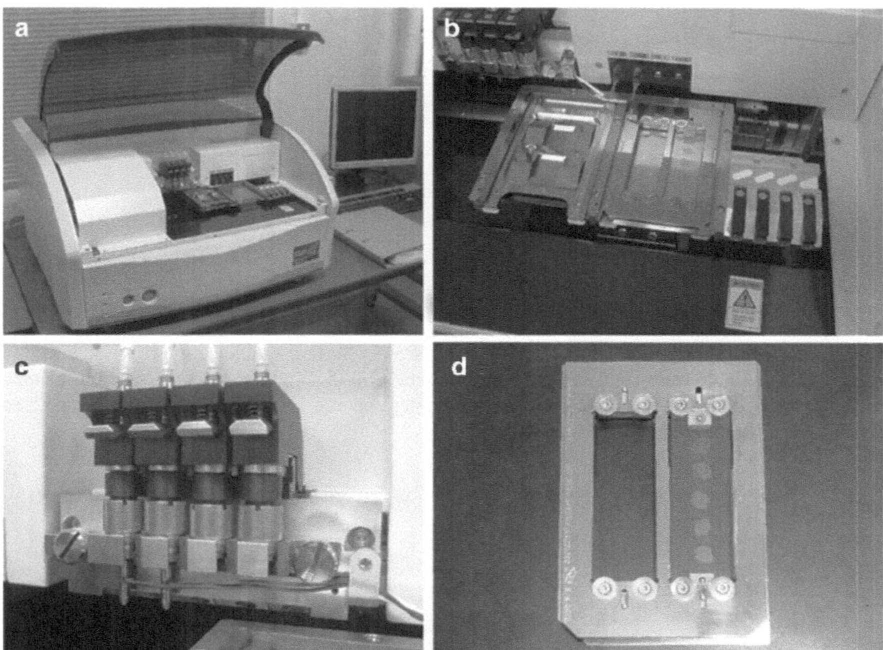

Fig. 6.6 ChiP-1000 and PC (a); stage (b); print heads (c). d Mouse cerebellum sections on an ITO-coated glass slide fixed on a target plate (Bruker)

When heavier digested peptides are measured using on-tissue digestion, the following steps are undertaken, in addition to steps 1–9 above:

16. Dispense a 100-pl volume of 15 mg ml^{-1} sinapinic acid (SA) in 6:4 acetonitrile / 0.5% TFA and deposit five drops (0.5 nl) in one spot.
17. Dispense at each position, at intervals of approximately 8 min, while the previously deposited spot dries.
18. Iterate steps (10) and (11) 20 times, to achieve a total drop volume of 10 nl at each position (Fig. 6.7).
19. Perform IMS measurement (Fig. 6.8).

6.4 Spray-Droplet Method

Pre-seeding of micro-crystals before applying the concentrated matrix solution is an effective procedure to produce more minute and homogeneous crystals on the tissue sections [1]. A combination of the two aforementioned methods can spectacularly improve the signal intensity and the signal-to-noise (S/N) ratios of mass spectra [2]. The so-called spray-droplet method first forms very fine matrix crystals

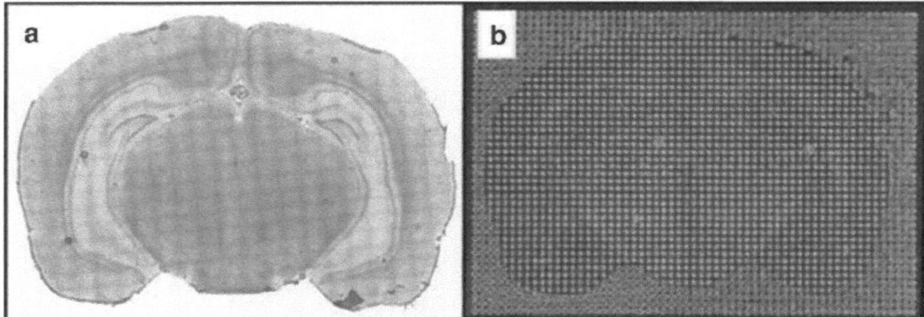

Fig. 6.7 **a** Hematoxylin and eosin (H&E) strain of a coronal rat brain section. **b** Tissue section used for on-tissue digestion and spotted with sinapinic acid (SA) matrix solution, which is printed as 2,500 spots consisting of 250 μm center-to-center spacing [10] (Reprinted from Groseclose et al., J Mass Spectrom 42(2):254–262.)

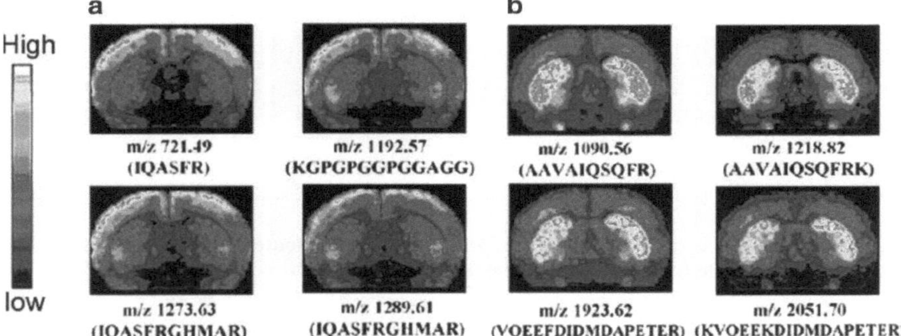

Fig. 6.8 Distribution of tryptic peptides generated from the digestion of the 7.5-kDa protein (**a**) and neurogranin and tryptic peptides generated from the digestion of the 6.7-kDa protein, PEP-19 (**b**), in rat brain coronal section. (Reprinted from Groseclose et al., J Mass Spectrom 42(2):254–262.)

on the surface of the tissue section by the spraying of low concentrated solution; they play the role of crystal nuclei for the subsequent droplet method, generating excellent homogeneous and minute matrix crystals in the spot (Figs. 6.9 and 6.10, top), improving signal intensity and S/N ratio. As a consequence, the number of detected peaks is increased (see Fig. 6.10, bottom). Below, we introduce this method.

6.4.1 Protocol

1. Place a mouse brain section mounted on an ITO glass slide in an airtight container with saturated water vapor at 37°C.
2. Spray 200 μl matrix solution (2 mg ml^{-1} SA in 1:1 acetonitrile and 0.1% TFA in water).

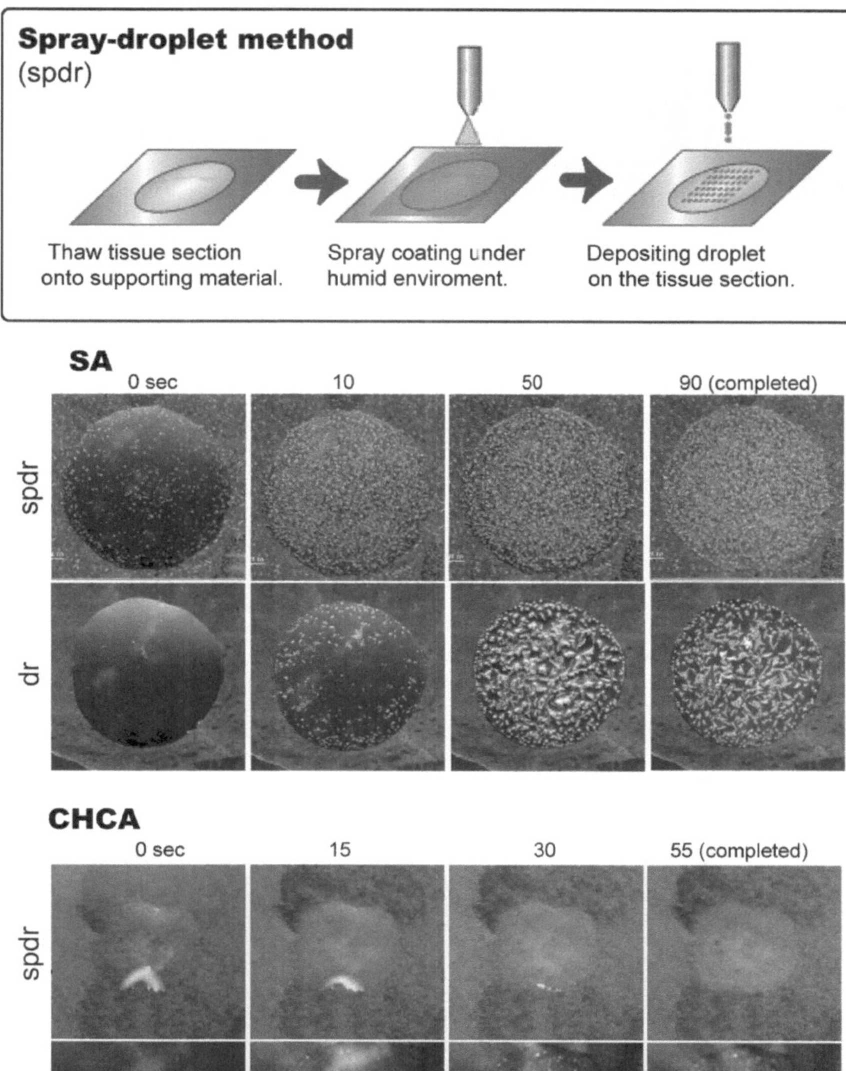

Fig. 6.9 *Top:* Schematic representation of spray-droplet method. *Bottom:* Time-lapse observation of matrix crystal formation. For both SA and α-cyano-4-hydroxy cinnamic acid (CHCA) matrix, the spray-droplet method (*spdr*) forms finer and more homogeneous crystals than those obtained by the droplet method (*dr*). (Partially reprinted from Sugiura et al., Anal. Chem 78(24):8227–8235.)

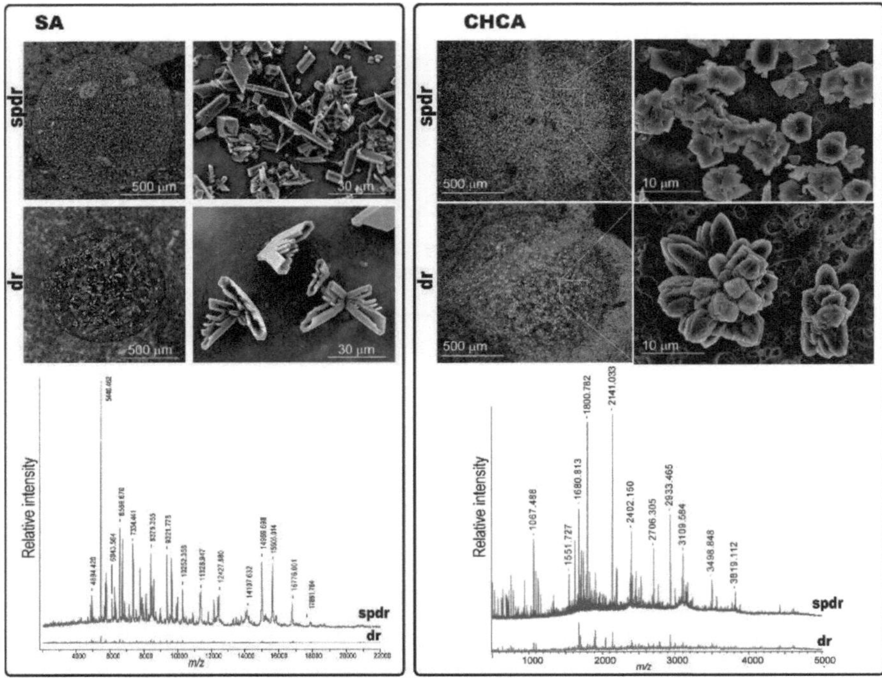

Fig. 6.10 Finer crystals formed by spray-droplet method improve spectrum quality. *Top panel:* Observation with a scanning electron microscope of matrix crystals with spray-droplet and droplet method. *Bottom panel:* Comparison of mass spectra obtained from rat brain. Optical observation of micro-spotted tissue sections employing spray-droplet and droplet method. (Partially reprinted from Sugiura et al., Anal Chem 78(24):8227–8235.)

3. Reseal the container and let the section dry for approximately 5 min.
4. Repeat steps (2) and (3) three more times (four times in all).
5. Place the section on the sample stage of the ChIP-1000.
6. Pour the matrix solution (8–10 mg ml^{-1} SA in 1:1 acetonitrile and 0.1% TFA in water) into the print head.
7. Designate the printing positions.
8. Dispense an 87-pl volume of the matrix solution and deposit 11 drops (approximately 1 nl) in one spot.
9. Dispense at each position, at intervals of approximately 10 min, while the previously deposited spot dries.
10. Iterate steps (8) and (9) until a total drop volume of 10 nl is obtained at each position.
11. Perform IMS measurement.

6.4.2 *Notes Regarding the Spray-Droplet Method*

- For best results, the matrix solution should be sprayed in a high-humidity environment, which generates optimal crystal nuclei. When doing so, spray matrix solution onto the section from an approximate distance of 20 cm and let the section become slightly wet.
- The compositions of matrix solution described above are for measuring proteins. When relatively small peptides are to be measured, we recommend the following: for generating matrix crystal nuclei with an air-brush, 8 mg ml^{-1} CHCA in 1:1 acetonitrile and 0.1% TFA; for dispensing matrix solution with a ChIP-1000, 6 mg ml^{-1} CHCA in 1:1 acetonitrile and 0.1% TFA. To prevent the nozzle of the ChIP-1000 from clogging, CHCA is to be used in concentrations <10 mg ml^{-1}.
- We recommend a drop volume of 87–100 pl, as this volume range provides stable behavior in the ChIP-1000 and reasonable IMS image resolution.

6.5 Matrix Sublimation Method

Finally, the matrix sublimation method, another novel methodology for matrix application, is briefly introduced. Recently, Hankin et al. reported that the commonly used matrices DHB and CHCA were found to undergo sublimation without decomposition under conditions of reduced pressure and elevated temperature [3]. This transition from solid to vapor phase was demonstrated as an effective way to apply the matrices onto tissue sections over a broad surface for IMS [3, 4]. Figure 6.11 shows a sublimator device for this method. The tissue slice on the MALDI plate insert is attached to the underside of the flat-bottom condenser and affixed by thermal conducting tape. Solid matrix was added to the bottom section of the apparatus, and the two pieces were assembled and connected to a direct drive vacuum pump. After several minutes at reduced pressure, the condenser was filled with cold water and heat was applied to the base of the sublimator. The sublimation of the solid matrix gradually progressed upon application of heat [3].

Sublimation of the matrix produced an even layer of small crystals across the sample plate (Fig. 6.12). This approach was found to yield highly reproducible results, eliminating much of the variance of the aforementioned parameters [4]. Furthermore, it is reported that ion image quality is enhanced by the microcrystalline morphology of the deposited matrix, increased purity of deposited matrix, and evenness of deposition [3, 4].

This solvent-free application method is so far applied for IMS of small molecules, mainly phospholipids. Compared to the spraying method, especially for small abundant molecules such as phosphatidylcholines, the sublimation method yielded a more intense and more even signal with fewer sodium adducts [3, 4].

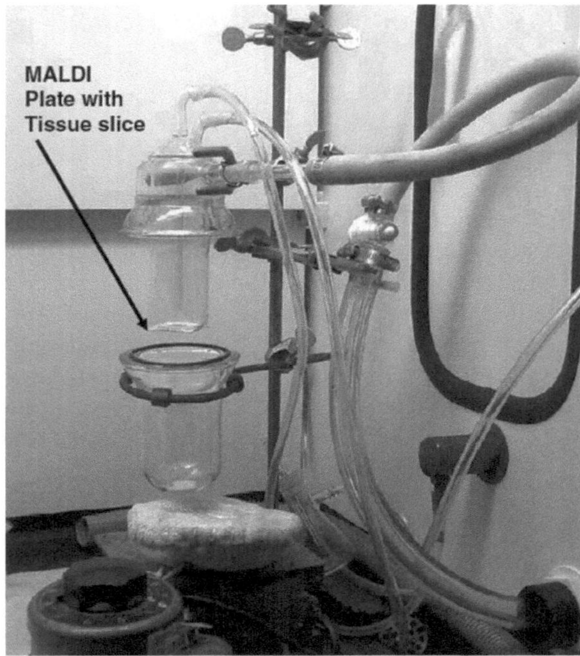

Fig. 6.11 Photograph of sublimation device used to apply matrix to tissue samples for MALDI-MSI experiments. The tissue slice (*arrow*) on the MALDI plate insert is attached to the underside of the condenser by thermal conducting tape. (Reprinted from Hankin et al., J Am Soc Mass Spectrom 18:1646–1652.)

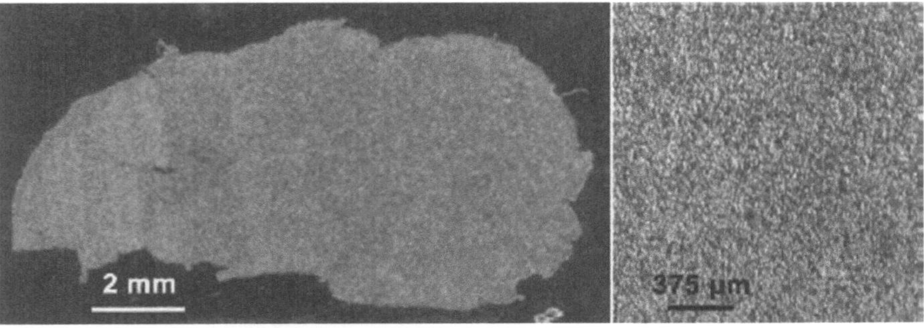

Fig. 6.12 A scanned optical image (*left*) of a dry-coated sagittal mouse brain section on a MALDI plate and an optical microscope image (*right*) of the DHB layer on the tissue at a magnification of ×4. (Reprinted from Puolitaival et al., J Am Soc Mass Spectrom 19:882–886.)

References

1. Aerni HR, Cornett DS, Caprioli RM (2006) Automated acoustic matrix deposition for MALDI sample preparation. Anal Chem 78:827–834
2. Sugiura Y, Shimma S, Setou M (2006) Two-step matrix application technique to improve ionization efficiency for matrix-assisted laser desorption/ionization in imaging mass spectrometry. Anal Chem 78:8227–8235
3. Hankin JA, Barkley RM, Murphy RC (2007) Sublimation as a method of matrix application for mass spectrometric imaging. J Am Soc Mass Spectrom 18:1646–1652
4. Puolitaival SM, Burnum KE, Cornett DS, et al. (2008) Solvent-free matrix dry-coating for MALDI imaging of phospholipids. J Am Soc Mass Spectrom 19:882–886
5. Agar NY, Yang HW, Carroll RS, et al. (2007) Matrix solution fixation: histology-compatible tissue preparation for MALDI mass spectrometry imaging. Anal Chem 79:7416–7423
6. Benabdellah F, Touboul D, Brunelle A, et al. (2009) In situ primary metabolites localization on a rat brain section by chemical mass spectrometry imaging. Anal. Chem., 81 (13), 5557–5560
7. Sugiura Y, Konishi Y, Zaima N, et al. (2009) Visualization of the cell-selective distribution of PUFA-containing phosphatidylcholines in mouse brain by imaging mass spectrometry. J. Lipid Res. 50: 1776–1788
8. Shimma S, Furuta M, Ichimura K, et al. (2006) A novel approach to in situ proteome analysis using chemical inkjet printing technology and MALDI-QIT-TOF tandem mass spectrometer. J Mass Spectrom Soc Jpn 54:133–140
9. Meistermann H, Norris JL, Aerni HR, et al. (2006) Biomarker discovery by imaging mass spectrometry: transthyretin is a biomarker for gentamicin-induced nephrotoxicity in rat. Mol Cell Proteomics 5:1876–1886
10. Groseclose MR, Andersson M, Hardesty WM, Caprioli RM (2007) Identification of proteins directly from tissues: in situ tryptic digestions coupled with imaging mass spectrometry. Journal of Mass Spectrometry 42(2):254–262

Chapter 7
Matrix Application with ImagePrep

Martin Schuerenberg and Soeren-Oliver Deininger

Abstract In this chapter, matrix application to tissue samples with the ImagePrep device is explained. The operating principle of vibrational vaporization of the matrix solvent and gravitational deposition of the droplets is presented. An important feature of the device is the control of the deposition process with an optical sensor; the principle of this detection is described. The chapter includes an introduction to the simple user interface with the adjustment of wetness, incubation time, and matrix thickness as intuitive parameters. A brief introduction to method development possibilities for detailed control over the matrix deposition process is given.

7.1 Introduction

ImagePrep is a stand-alone preparation device used to apply matrix to tissue samples for MALDI imaging (Fig. 7.1).

A fine mist of matrix droplets is generated through the vibration of a pinholed metal sheet. The fine matrix droplets are allowed to sink into the tissue by gravitational force alone and to incubate under controlled conditions. The ImagePrep is operated via an integrated touch-panel screen. The preparation is in real time; it is controlled by an internal optical sensor to achieve the highest possible reproducibility. Because of its unique design, the ImagePrep produces small droplets with a 25 µm diameter; additionally, they are produced simultaneously, to ensure efficient incubation. This capability leads concurrently to a high lateral resolution and good mass spectra (Fig. 7.2), which are mutually exclusive aims when other sample preparation devices are used. The instrument allows two levels of operation: a simple user interface and a method development graphic user interface (GUI) for expert users.

M. Schuerenberg and S.-O. Deininger (✉)
Bruker Daltonik GmbH, Fahrenheitstrasse 428359, Bremen, Germany
e-mail: Soeren-Oliver.Deininger@bdal.de

M. Setou (ed.), *Imaging Mass Spectrometry: Protocols for Mass Microscopy*,
DOI 10.1007/978-4-431-09425-8_7, © Springer 2010

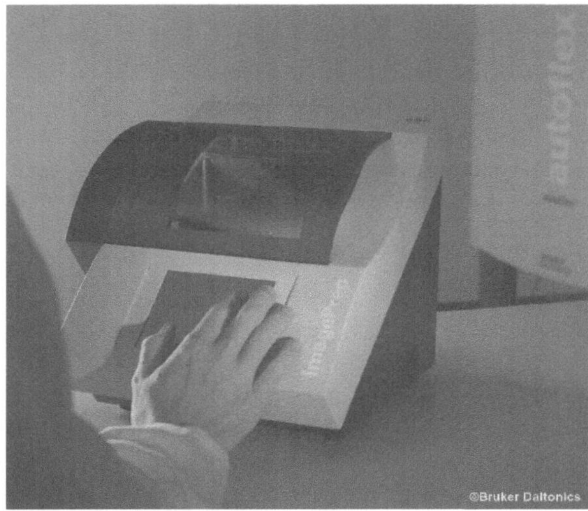

Fig. 7.1 The ImagePrep preparation device

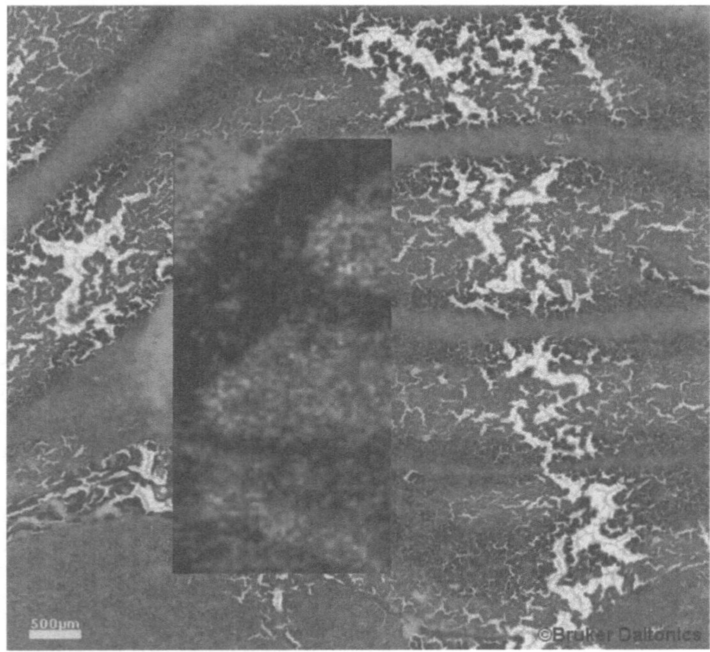

Fig. 7.2 Correlation between histological staining and matrix-assisted laser desorption/ionization (MALDI) image (*insert*). The MALDI image shows the distribution of two different phospholipids at a 25-μm spatial resolution

7.2 The Simple GUI

This mode is considered the standard mode of operation. It allows a true push-button operation and does not require a detailed understanding of the underlying technology. The user must choose a predefined method, according to the matrix he or she has chosen for the imaging experiment. Upon choosing a method, an information screen with the correct matrix concentration and solvent composition is shown on the screen. After filling the reservoir with matrix, preparation can be started immediately. The matrix application can be customized by changing three intuitive parameters: incubation time, matrix layer thickness, and wetness during preparation. Because these parameters may need to be optimized for different tissue types, customized settings can be saved for later use.

7.3 The Operation

To build up a sufficient matrix layer thickness without the lateral delocalization of analytes, it is necessary to apply the matrix in multiple cycles. Each cycle consists of three phases. In the first phase, the spray is generated. During the second phase, the matrix is incubated on the tissue section. In the third phase, a soft nitrogen flow is used to dry the sample and therefore ensure a consistent incubation time between cycles. The matrix preparation is ended after the desired thickness of matrix, as defined by the method, is reached. After this, the sample can be directly measured. The instrument can be conveniently cleaned by running a cleaning program and wiping the Teflon-coated spray chamber.

7.4 The Optical Sensor

The heart of the ImagePrep is an optical sensor that enables the real-time monitoring of a scattered light signal. The sensor and the light source are located under the sample, which needs to be mounted onto a glass slide equipped with an electrically conductive ITO coating. To understand the principle of the optical sensor (Fig. 7.3), it is best to image a sample that is already covered with matrix crystals. These crystals will scatter the light to a certain extent. When the spray is started, the surface of the sample becomes wet; because the liquid droplets smooth out the craggy crystals (i.e., refractive index matching), the matrix layer becomes more transparent. Therefore, less light is scattered and the scattered light signal decreases. The decrease of the scattered light signal is a direct measurement of the wetness of the sample. As the sample dries, the scattered light signal increases again and the drying rate can be directly monitored by the scattered light signal. If the drying is

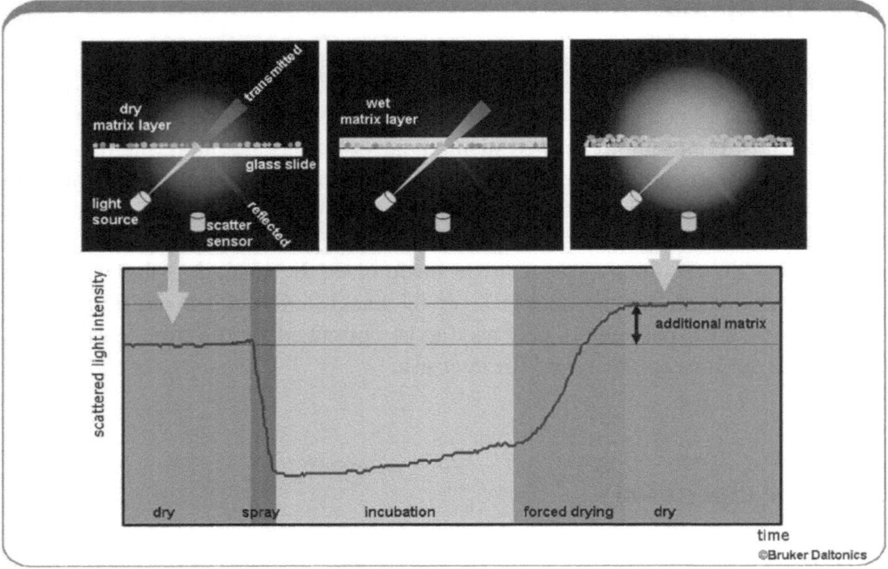

Fig. 7.3 Function of the optical sensor

completed, additional matrix crystals are formed, so the scattered light signal will be higher than in the beginning. When the sample is dry, the scattered light signal will therefore serve as a direct measure for the thickness of the matrix layer.

During a typical preparation, the ImagePrep will not dry the sample completely after each matrix application but instead applies the next matrix layer while the sample is still wet, which ensures a maximum extraction efficiency and good spectra quality in the imaging experiment. After a certain number of spray cycles, a complete drying is performed, to estimate the absolute matrix layer thickness. The next spray cycle is then applied until a defined wetness is achieved.

7.5 The Method Development GUI

For the routine use of ImagePrep with standard matrices such as sinapinic acid (SA), 2,5-dihydroxybenzoic acid (DHB), or cyano-4-hydroxy cinnamic acid (CHCA), no understanding of the internal workflow of the instrument operation is necessary. Users can choose special matrices or create their own methods, via the method development GUI. Because the operation of ImagePrep is controlled by the scattered light sensor, it is necessary to have access to the scattered light output of the sensor during method development. An analog/digital converter and suitable software are provided with ImagePrep, allowing access to the scattered light curves. Through the method development GUI, every relevant sample preparation parameter is accessible, and the entire preparation process can be divided into different phases. For each phase, the experimenter can decide whether a fixed number of spray cycles shall be used or if the matrix layer thickness will be controlled via the sensor. For each spray cycle within a phase, the experimenter can define how long the spray shall be on and whether this spray-on time will be controlled by the sensor; he or she can next decide how long the sample will be incubated before the nitrogen flow starts, as well as at which level of dryness the next spray cycle will be applied. There are more parameters that can be controlled. A detailed manual for method development is provided with ImagePrep.

Part IV
Instrumental Operation for IMS

Chapter 8
Method of Operating AXIMA-QIT as Imaging Instrument

Takahiro Hayasaka

Abstract AXIMA-QIT (Shimadzu Co.) is a device to be used for MALDI, QIT, and TOF-MS, and it is used for IMS with additional use of an original converter. The technique enables us to visualize the distribution and the relative volume of biomolecules on a tissue slice. Especially, QIT has a function of trapping an ionized biomolecule at high resolution, and the ions fragmented in QIT give us the structural information of the biomolecule. This procedure can be infinitely iterated in theory (MS^n analysis). Therefore, AXIMA-QIT can identify biomolecules with a complicated structure as well as visualize the distribution of biomolecules on a tissue slice. In this chapter, the protocol for IMS and identification of biomolecules by MS^n analysis with AXIMA-QIT are introduced.

8.1 Outline of AXIMA-QIT-TOF-MS

The AXIMA-QIT-TOF-MS is a device that can be used to perform imaging mass spectrometry (IMS) (Fig. 8.1). This device comprises a matrix-assisted laser desorption/ionization (MALDI) ion source, to ionize the sample; a quadrupole ion trap (QIT), to trap the ions; and time-of-flight (TOF)-MS, to detect the mass using TOF separation. In particular, QIT has a high ion selectivity and a function that enables, theoretically, the infinite fragmentation of ions (i.e., MS^n analysis). Detailed structural analysis is required, as well as the imaging of the localization of biological molecules in advanced IMS. AXIMA-QIT-TOF-MS is a device that can fulfill this demand with high sensitivity, and in concert with the MS^n analyses.

This section explains the principle underlying QIT. QIT comprises a ring electrode and an end-cap electrode. When an ion is introduced into QIT, the ion can be trapped within; the ion is stabilized by combining the high-frequency voltage on the ring electrode and direct current voltages on the end-cap electrode. Introducing

T. Hayasaka(✉)
Department of Molecular Anatomy, Hamamatsu University School of Medicine,
1-20-1 Handayama, Higashi-ku, Hamamatsu, Shizuoka 431-3192, Japan
e-mail: thaya@hama-med.ac.jp

M. Setou (ed.), *Imaging Mass Spectrometry: Protocols for Mass Microscopy*,
DOI 10.1007/978-4-431-09425-8_8, © Springer 2010

Fig. 8.1 AXIMA-QIT IMS system

helium gas decreases the kinetic energy of the ion in QIT. The ion is then discharged to TOF-MS, where the MS spectrum is measured.

In MS/MS analysis, the movement of ions selected in a particular mass range is stabilized by adjusting the combination of voltages applied to the ring electrode and the end-cap electrode. Other ions are discharged from QIT by making the movement unstable in a contrasting manner. As a result, the ions of specific mass and helium gas – namely, the cooling gas that decreases the kinetic energy of the ion to be trapped – remain in QIT. When argon gas is introduced into QIT, the ion is fragmented by making it collide with the gas. The MS/MS spectrum is measured by releasing the fragment ions into TOF-MS. In AXIMA-QIT-TOF-MS, the MS^n spectrum can be measured by repeating the process of trapping and fragmenting the ions in QIT, as already mentioned.

8.2 IMS by AXIMA-QIT-TOF-MS

In this section, the process of putting the sample onto the MS target plate and performing a raster-scan measurement in IMS using AXIMA-QIT-TOF-MS is explained. The measurement is completed by acquiring the information at the position of the sample on the MS target plate by ChIP-1000, shifting the information to AXIMA-QIT-TOF-MS, and executing the raster scan.

8.2.1 Acquisition of Position Information by ChIP-1000

The ChIP-1000 software is opened on the computer where it has been installed, and the operator starts the program using the user ID and password provided for the software:

- The ChIP-1000 window is displayed when the software starts. The user ID is displayed on the left side. For example, the user ID "a" is used for demonstration purposes.
- The [Routine] option is selected in the [Operate] – [Operation Mode]. [Routine] is selected if there is a check in [Area Print]. The sample window starts when [Operate] – [Sample] is clicked.
- By this stage, the MS target plate have been inserted into the ChIP-1000 (Fig. 8.2). It is inserted in the position of 'Sample 1."
- The scanning of the MS target plate is begun by clicking "Image Acquisition," after the options for "Display Status" and "Sample 1" are check-marked.
- If the MS target plate is set up in the position of "Sample 2," the option for "Sample 2" also needs to be check-marked.
- The software takes approximately 20 s to scan the sample image, and the status of the scanning is displayed.
- A new "Image Name 1" window opens when the scanning ends.
- The scanning data are preserved by clicking the "Save" button at the lower right. The preservation of scanning data is completed by clicking the "OK" button, after selecting "Project Name" and "Experiment Name" and inputting the "Sample Name." As an example, "Test Sample" was input as the "Sample Name."

Fig. 8.2

- Moreover, the "New" button is be clicked when "Project Name" and "Experiment Name" are newly created. An arbitrary name is saved if no names are specifically input.
- The "Save a Sample" window then closes and the next screen is visible. When the scanning data is normally preserved, the name of the window changes from "Image Name 1" to "Test sample."
- Next, the process of acquiring information at the position of the sample on the MS target plate begins. First, the "Zoom" button is clicked in the "Test Sample" window, and the "Zoomed Image" window opens.
- The entire image and the expanded image of the MS target plate are displayed on the screen.
- The position of the expanded image corresponds to an orange frame on the entire image.
- The area that is thus magnified can be changed by moving this orange frame while holding down the left-click button of the mouse.
- A red cross appears on the expanded image when the "Add Manually" button is clicked. An orange "×" marker is created at the part of the cursor where the left-click button is.
- Numbers are recorded in the order in which they are created (Fig. 8.3).
- AXIMA-QIT-TOF-MS provides a function whereby the cursor is automatically moved to the position of the "x" marker; therefore, the operator marks the spot

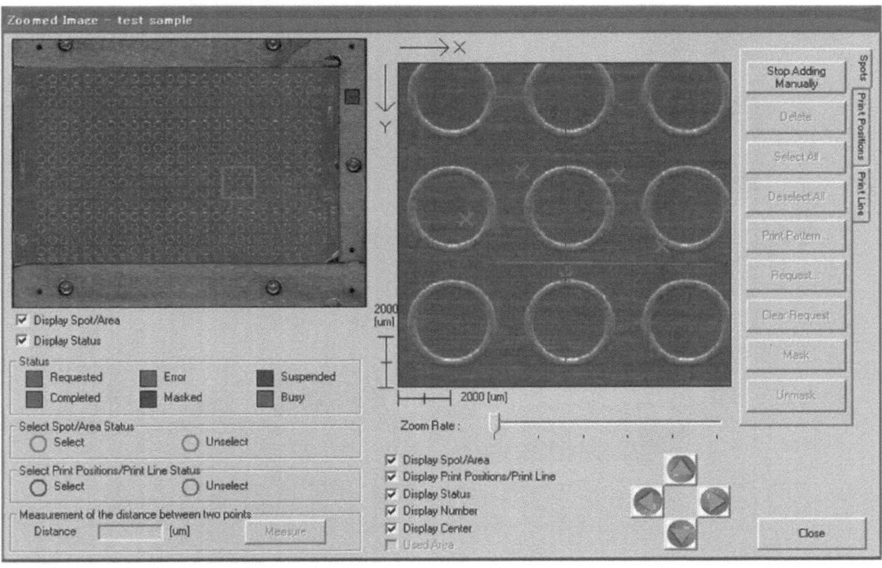

Fig. 8.3

where standard samples for calibration are deposited and the measurement areas in IMS.

- When completed, upon clicking the "Close" button, the window shuts.
- When the "Print Pattern" button is clicked, the "Select" window opens. The "OK" button is clicked after checking "Print Position."
- When the "Print Pattern" window is closed, the markers are in a nonselective state. By clicking on the "Select All" button, the markers become the selective state.
- The "Select Analyzer" window opens when the "To Analyzer" button is clicked. The "OK" button is clicked with the "To AXIMA" option check-marked.
- Information regarding the sample position that is to be transferred to AXIMA-QIT-TOF-MS is preserved as a CSV file. For example, the file name was pre-served as "Test Sample" on the computer desktop; a CSV file with the name "Test Sample" is observed there (Fig. 8.4).
- When ChIP-1000 completely acquires the information regarding the sample position on the MS target plate, we shut down the program by clicking the "Close" button in "Test Sample."
- Then, the "Close" button in the "Sample" window is clicked and this window closes. [Operate] – [Exit] is clicked.
- The "OK" button is clicked twice after receiving two messages of confirmation for ending the program.

Fig. 8.4

8.2.2   Settings in AXIMA-QIT-TOF-MS

AXIMA-QIT-TOF-MS can be operated by the software package Kompact, if it is installed in the connected PC:

- First, the operator opens the loading slot of the sample, to introduce the MS target plate into AXIMA-QIT-TOF-MS.
- After it is confirmed that the "Mode" of the Acquisition window is "Standby," the "Open Door" button is clicked. Because the loading slot is opened after a while, the MS target plate is introduced.
- The loading slot shuts when the "Close Door" button is clicked; at that point, the sample chamber begins to be vacuumed.
- In AXIMA-QIT-TOF-MS, the first operation is limited until Gauge 1 reaches a value of 2.0E-3. Meanwhile, the operator may load the position information of the sample made using ChIP-1000. The CSV file ("Test Sample.csv") is copied to the PC connected AXIMA-QIT-TOF-MS.
- The "Open" window in the "Slide" tab in Acquisition is opened and the "Load" button is clicked. When "Look in" is specified for a hierarchy where the CSV file is placed and "File of Type" is made "All Files," the CSV file is displayed.
- When "Test Sample.csv" is specified and the "Open" button is clicked, the position information with "×" markers is read.
- The value of Gauge 1 decreases to 2.0E-3 or less as the procedure progresses. The dotted line of the pink color is displayed in the "Instrument Status" window in the lower right of the PC screen, upon clicking the "Operate" button (Fig. 8.5).
- It returns to the "Slide" tab again; the "Align Plate" button is then clicked to correct the position of the MS target plate in AXIMA-QIT-TOF MS.

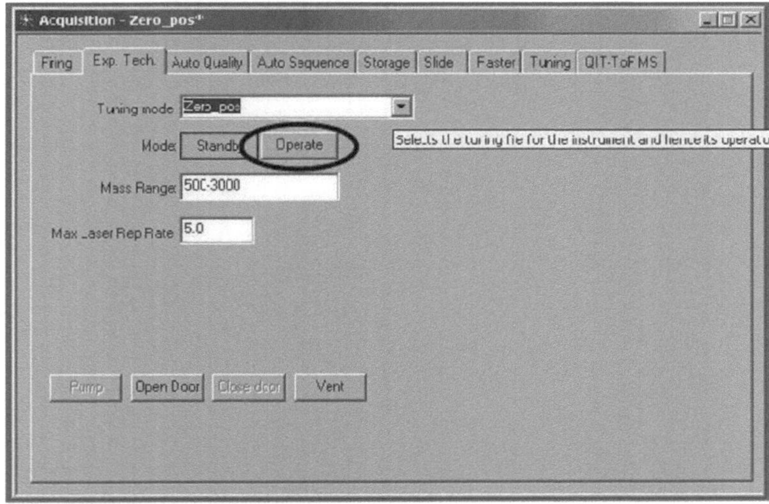

Fig. 8.5

- After the camera mark of "Align Sample Plate" window is clicked, a real-time image of the MS target plate is displayed.
- When the "Firing" tab is clicked, the pattern diagrams of the MS target plate are painted out with yellow and displayed. The black frame shown under the left side of the MS target plate pattern diagrams corresponds to the position of the CCD camera.
- The CCD camera moves to the position of ref 1 after activating "Ref 1" on the "Align Sample Plate" and having pressed the right arrow (→) button.
- Moreover, it is confirmed that the marker created with ChIP-1000 is displayed on the MS target plate pattern diagrams.
- The CCD camera moves to the position indicated by the on-screen cursor; it is adjusted so that a red cross becomes the center of the hole in the MS target plate (Fig. 8.6).
- By clicking on the "Set Reference" button of the "Align Sample Plate," the setting is completed.
- When the "OK" button is clicked after repeating the same operation for "ref 3," the correction by three points is completed.
- The setting is preserved by clicking the "Apply" button on the "Slide" tab at the end.

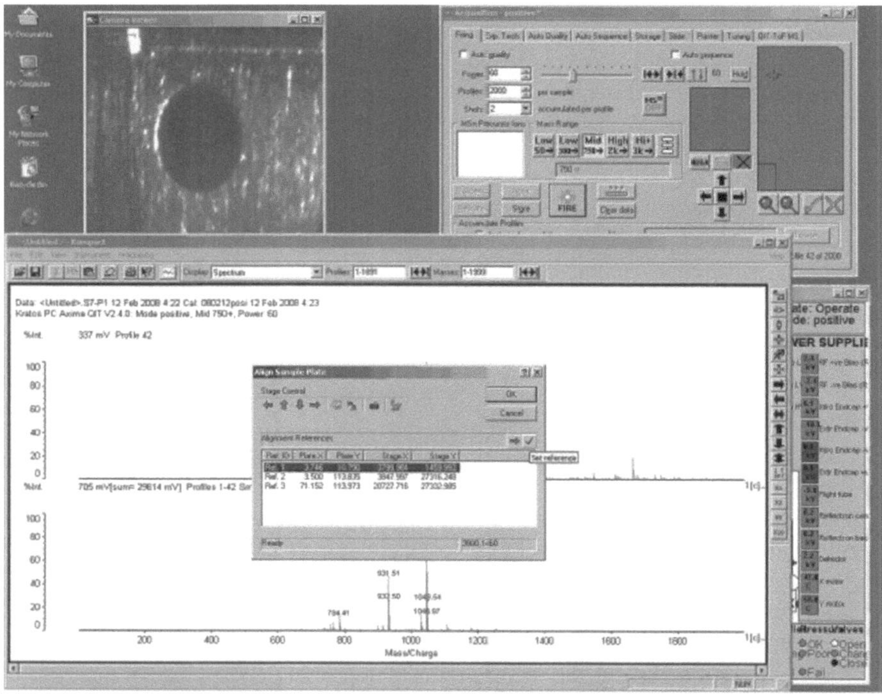

Fig. 8.6

8.2.3   Achieving Settings in AXIMA-QIT-TOF-MS

The positive or negative mode is selected by clicking the "Positive" or "Negative" button on the "Exp. Tech." tab, respectively, as required. It is important to make this selection in the following order, to avoid any electrical discharge caused by a rapid load.

Positive
↓↑
zero posi
↓↑
zero nega
↓↑
negative

- The black frame in the pattern diagrams of the MS target plate is moved to where the operator wants to measure by using the mouse's left-click button. It was moved to the area where the position of the sample was indicated by the markers.
- A red cross in the CCD image shows the position where the laser is currently irradiating the sample. "Mass Range" is selected, according to the mass range for the biomolecules of interest. For example, it has been set to "Mid 750→" to measure a large range of lipids that exist in m/z 750–800.
- [Kompact] – [Instrument] – [QIT-TOF Experiment] is clicked, and the "Experiment" window opens when the password is entered. The "Open" window is displayed when [File] – [Open] is selected, and "Yes" is clicked for "Warning."
- The setting of the best measurement in the equipment is loaded by selecting "PosMidMass" among the prepared configuration files.

8.2.4   Calibration

It is necessary to calibrate the condition at each measurement. Angiotensin II and its fragmented molecules were used here:

- The power setting is needed to be changed according to the state of each model. In this case, the power was set to 70 to produce a sufficient quantity of fragmented molecules.
- [Processing] – [Calibration] is clicked to calibrate through the use of the detected signal.
- The numerical value is displayed in the column below "Mass," when the list reference is opened and a target numerical value (m/z 1,046, in the present case) is clicked.
- If the cursor is matched to the peak at m/z 1,046 and the center of the peak is clicked with the mouse wheel, the measurement values are displayed in "Current Mass."
- It is fixed with an upper and lower arrow button (↑↓).
- The value of each detected peak is corrected by similar processing for m/z 931, m/z 784, and so on, by clicking the "Calibrate" button (Fig. 8.7).
- Finally, the calibration parameter is preserved by clicking "Save."

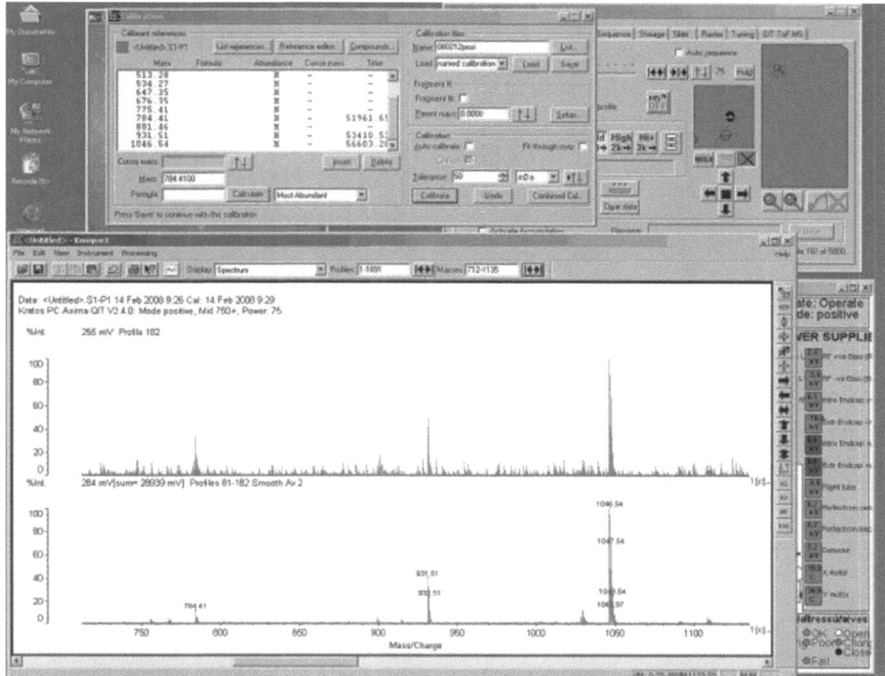

Fig. 8.7

8.2.5   Setting of Measurement Area

In the tissue slice projected onto the CCD camera, the size can be determined by referencing the scale bar shown in the screen. However, the true shape is not visible, because the lens of the CCD camera does not show the tissue slice from directly above. Therefore, the IMS measurement area needs to be simulated:

- First, the measured area is decided. The area is assumed to be 1,000 nm × 1,000 nm, and so the value of 1,000 is entered into each of the "Width" and "Height" blanks in the "Raster" tab (Fig. 8.8).
- When "Calculate Raster" is clicked after the "Based on Spacing" option is check-marked and a value of 1,000 is input into the "Spacing" blank, "Points" is automatically calculated as having a value of 4.
- These four points represent the four corners of 1,000 nm × 1,000 nm. Next, the values of 0, 4, and 2 are input into the "Power," "Profiles," and "Shots" blanks in the "Firing" tab, respectively.
- When the "FIRE" button is clicked after check-marking the button of a right and left arrows, it begins to irradiate the four points with the laser.
- It is then necessary to adjust "Profiles" to the same value as "Points" and set "Shots" to 2.

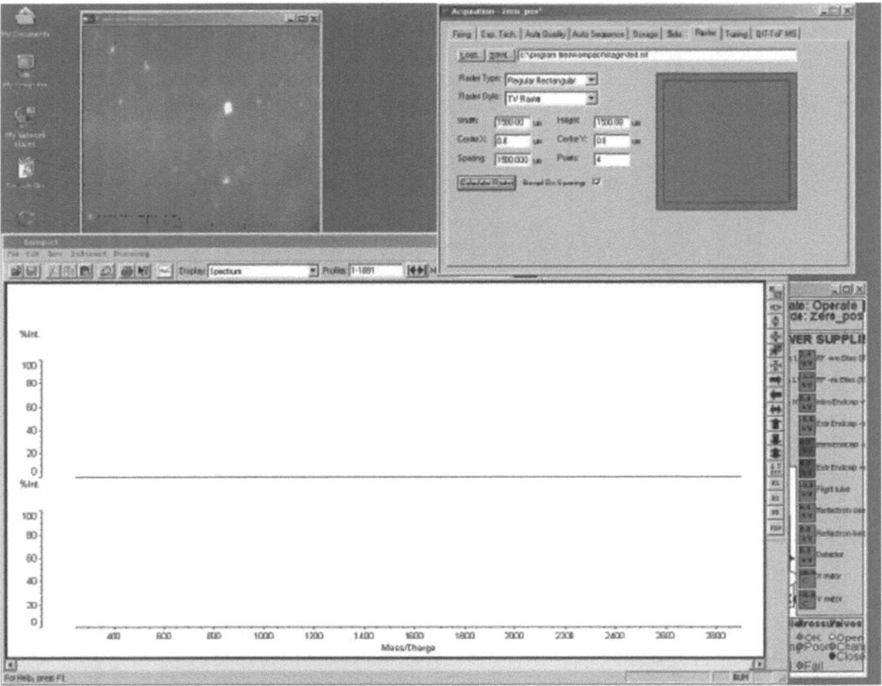

Fig. 8.8

- The "Shots" means the multiplication frequency.
- If the measurement area is 1,500 nm × 1,000 nm, "Points" is calculated as 12 by setting "Spacing" to 500, which is the greatest common factor of 1,500 and 1,000.
- The simulation of the measurement area is completed in a short time when the greatest common factor is used.

8.2.6 IMS Measurement

- The above-mentioned steps almost complete the setting of the IMS apparatus, as the laser power and shot numbers are determined. Here, when "Spacing" was set at 50 in the area of 1,000 μm × 1,000 μm, the value of "Points" was 441.
- Next, the setting is preserved by selecting "All" from "Store Profiles" in the "Storage" tab and clicking "Apply."
- After the numerical value of "Points" is synchronized with "Profiles" and the power and shot values are set, "FIRE" is clicked (Fig. 8.9).
- When the settings are 441 profiles and 20 shots, the measurement time is approximately 30 min. The ion image at m/z 798.79 in a mouse retina section is reconstructed using BioMap software (Fig. 8.10).

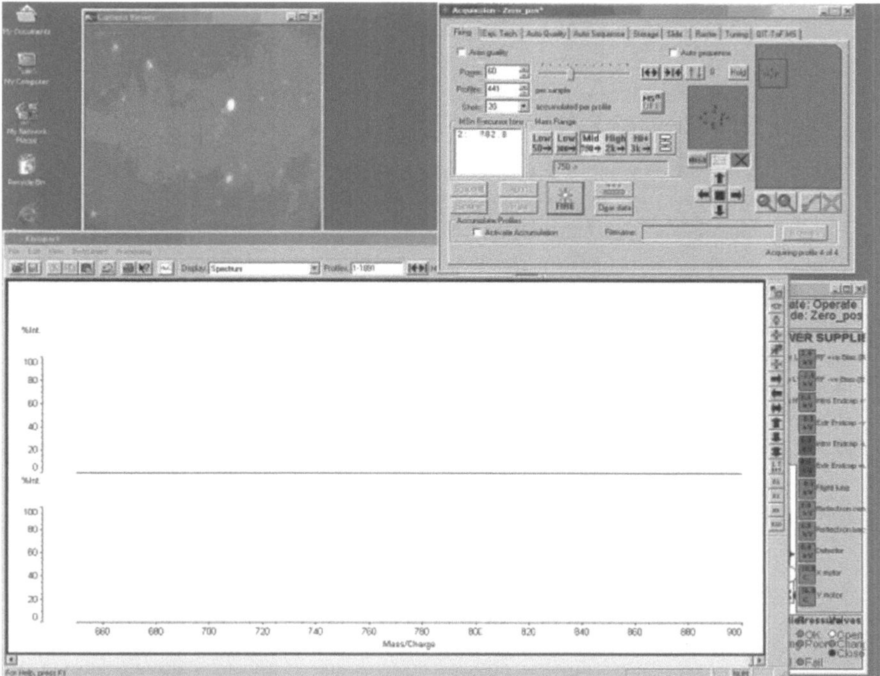

Fig. 8.9

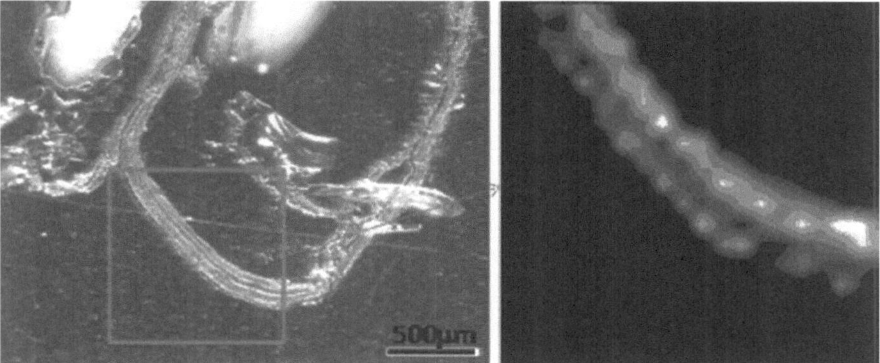

Fig. 8.10

8.3   Tandem MS by AXIMA-QIT-TOF-MS

The procedure for performing tandem MS analysis for m/z 798.79 is explained below.

8.3.1 Selection of Peak

Among the molecules ionized in the MALDI ion chamber, it is necessary to trap
only the target molecule by QIT:

- The molecule of interest is the value of m/z 798.79; therefore, 798.79 is entered
  as the "Precursor Ion" in the "QIT-ToF MS" tab, and the setting is preserved by
  clicking "Apply."
- When the "MS^n" button from the "Firing" tab is clicked, the button is reversed
  from "OFF" (Red) to "ON" (Green).
- When "FIRE" is clicked, did the detected peak show a value of m/z 798.79? The
  error margin of the numerical value should be approximately ±0.05.
- Here, the value of the detected peak was m/z 798.77; therefore, the settings for
  selecting the peak can be considered excellent (Fig. 8.11).
- If the setting is not good, the signal intensity of the selected peak may be small
  or other peaks are strongly detected. In such cases, the value of the "Precursor
  Ion" or the position of the laser irradiation is changed.

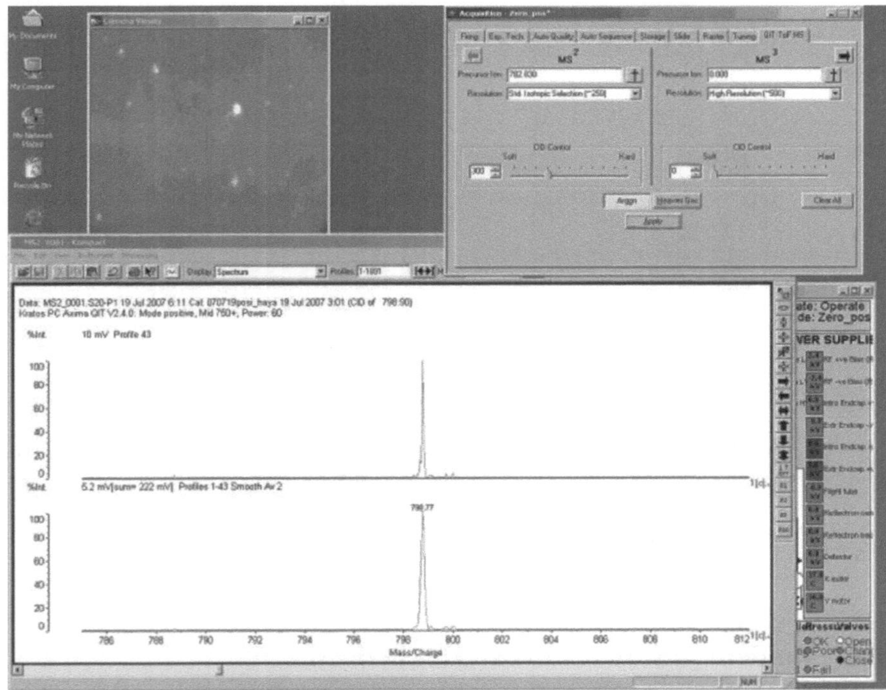

Fig. 8.11

8.3.2 MS² Measurement

The value of "CID Control" in the "QIT-ToF MS" tab is set when the selective condition of the peak is decided. This, in turn, depends upon the state of the equipment and the structure of the biomolecule. It is possible to fragment the ion sufficiently by setting a value of approximately 300 when the targeted molecule is a lipid:

- The setting is also preserved by clicking "Apply."
- The measurement begins by clicking the "FIRE" button in the "Firing" Tab.
- The peaks detected by the measurement were m/z 739.65, m/z 674.48, and m/z 615.33 (Fig. 8.12); these fragment peaks are 59, 124, and 183 units smaller, respectively, than m/z 798.79, which is the precursor ion.
- These fragment ions suggest that the biomolecule is either phosphatidylcholine or sphingomyelin; because sphingomyelin provides the precursor ion of odd values, the biomolecule is presumed to be phosphatidylcholine, given that the precursor ion is an even number value.

An MS/MS/MS analysis enables the further specification of the lipid. For instance, m/z 674.48 obtained from m/z 798.79 is assumed to be a precursor ion. The value of 674.48 is input as the "Precursor Ion" of MS³; it can be fragmented by enlarging the

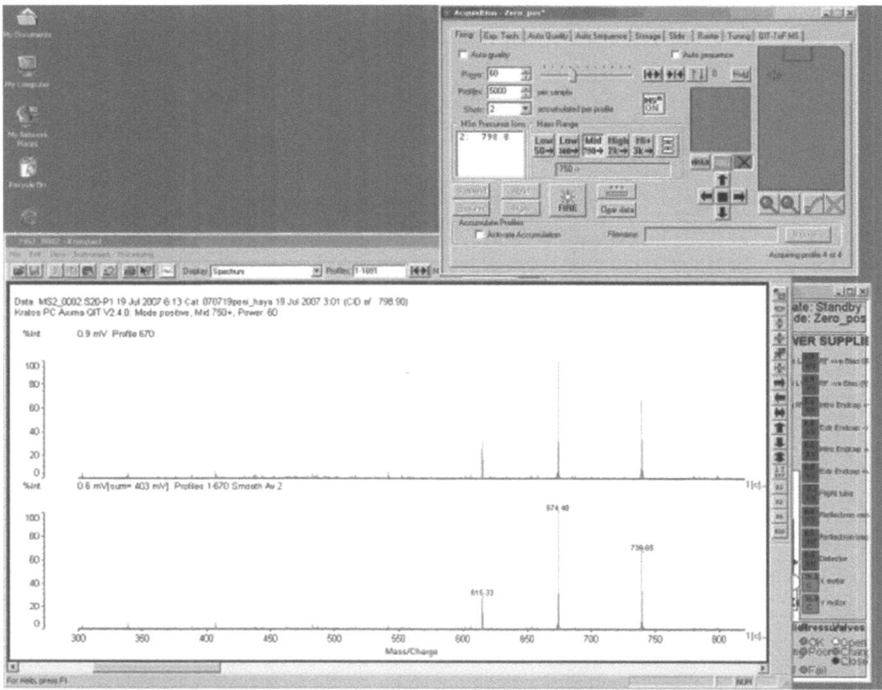

Fig. 8.12

value of "CID Control" of MS^3 to more than the setting when MS/MS analysis is performed. When fragment peak indicates a value that is smaller than the precursor ion by 22 or 38 units, the peaks were the result of the addition of Na^+ or K^+, respectively. For instance, it is understood that a fragment peak with a value that is 38 units smaller than m/z 674.48 is a K^+ addition. As a result, the retrieved molecule is a PC of m/z 760.53 or m/z 760.59, corresponding to the K^+ adduct. To acquire such detailed identification, it is necessary to detect the peaks of fatty acids from the lipid.

8.3.3 Identification of Peptide

As an example, a peak of m/z 2,141.10 and its fragment ions are measured here. It is possible to identify the peptide from the fragment peaks using the "Mascot Search" database (http://www.matrixscience.com/search_form_select.html). To export the fragment peak list as an ASCII file, [Export] – [File] – [ASCII] is clicked. The "Export ASCII" window is opened, and "Save As" is clicked after setting the following parameters: File format: PC; Column: 1; Delimiter: space; Decimal place: 2; Export: peaks; Format: mass, and intensity (Fig. 8.13).

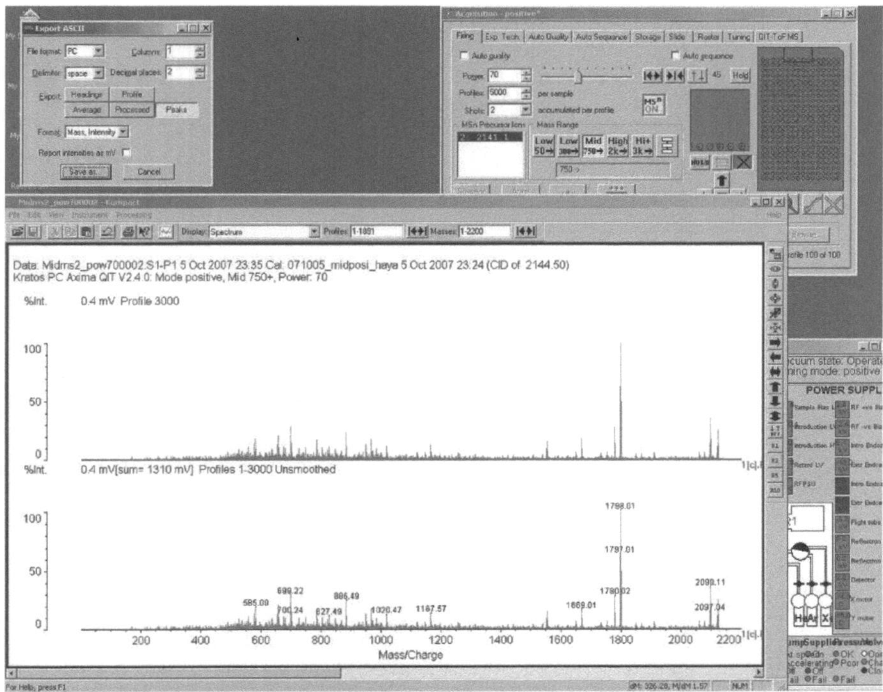

Fig. 8.13

The ASCII file is opened, and the commands of "begin ions," "end ions," and "pepmass = 2141.10" are input as follows.

(example)

begin ions

Pepmass = 2141.10

***.**

***.**

end ions

"Mascot Search" is opened in the Internet browser, and "MS/MS Ion Search" is selected. Peptide charge: "+1" and Instrument: "MALDI-QIT-TOF" are selected, the ASCII file created as a Data file is input. When "Start Search" is clicked, the retrieval is initiated (Fig. 8.14). After a while, the retrieval result showed that the peptide with the peak of m/z 2,141.10 is a myelin basic protein.

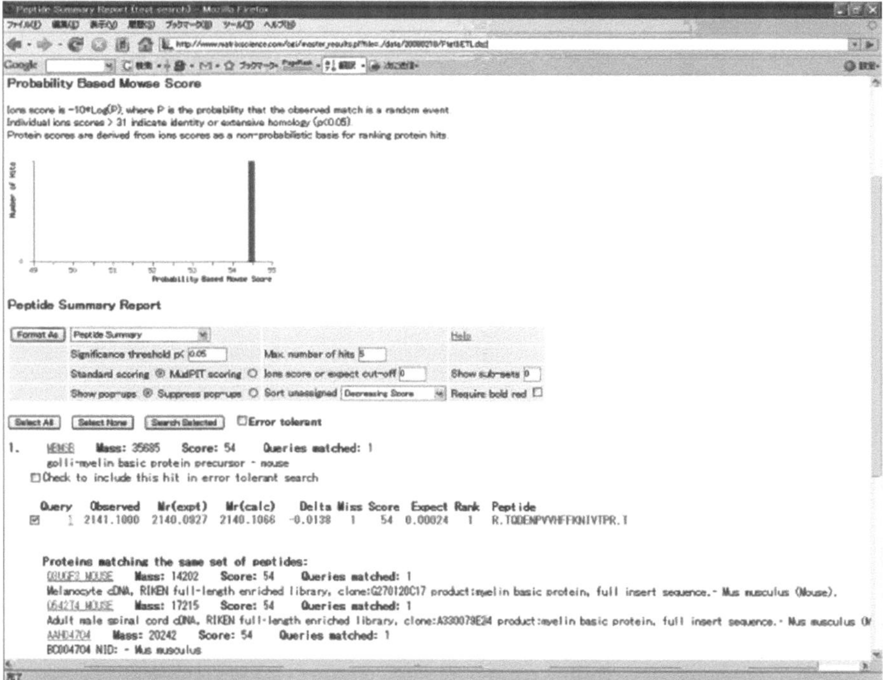

Fig. 8.14

Part V
Analysis of Measurement Data

Chapter 9
Ion Image Reconstruction Using BioMap Software

Naofumi Hosokawa, Yuki Sugiura, and Mitsutoshi Setou

Abstract This chapter is devoted to introducing the basic use of the BioMap for IMS data analysis. BioMap (free software; http://www.maldi-msi.org) is widely used to process mass spectral data for creation of two-dimensional coordinates of the surface versus intensity with excellent usability. In this chapter, we introduce the following methods: (1) how to set up the BioMap software, (2) the method of ion image reconstruction, and (3) data analyses of the mass spectra among the multiple regions of interest (ROI) and further data analyses; for instance, the differential analysis of mass spectra between the ROIs.

9.1 Introduction

For IMS experimenters, up to now, several types of software have been developed for data acquisition and image reconstruction by manufacturers and academic researchers (Table 9.1) [1]. Now, in various environments, it is possible to use a mass spectrometer as an imaging instrument and reconstruct a two-dimensional ion image from an acquired array of mass spectra. Among the data analysis software, BioMap (free software; http://www.maldi-msi.org) is a useful tool widely used by many researchers and enterprises because of its excellent utility. BioMap provides a set of analysis tools such as display of an ion distribution image of interest, review of spectra behind each pixel, definition of regions of interest (ROI) for mass spectral

N. Hosokawa and Y. Sugiura
Department of Bioscience and Biotechnology, Tokyo Institute of Technology,
4259 Nagatsuta-cho, Midori-ku, Yokohama, Kanagawa 226-8501, Japan

N. Hosokawa, Y. Sugiura and M. Setou
Mitsubishi Kagaku Institute of Life Sciences, 11 Minamiooya, Machida, Tokyo 194-8511, Japan

Y. Sugiura and M. Setou (⊠)
Department of Molecular Anatomy, Hamamatsu University School of Medicine,
1-20-1 Handayama, Higashi-ku, Hamamatsu, Shizuoka 431-3192, Japan
e-mail: setou@hama-med.ac.jp

M. Setou (ed.), *Imaging Mass Spectrometry*: *Protocols for Mass Microscopy*,
DOI 10.1007/978-4-431-09425-8_9, © Springer 2010

Table 9.1 Imaging mass spectrometry (IMS) device and software

	Instrument	Type	Software	Data format
Applied Biosystems/MDS Sciex	QSTAR® Elite System	QqTOF	oMALDI Server 5.0 with Imaging. BioMap	Wiff, Analyze
	4800 MALDI TOF/ TOF™ Analyzer	TOF–TOF	4800 Img. BioMap	Analyze
	Voyager DE™ PRO Workstation	TOF	MMSIT, BioMap	Analyze
Bruker	Ultrafex III	TOF–TOF	flexImaging, CreateTarget, AnalyzeThis! BioMap	Xmass, XML
Shimadzu	AXIMA-TOF^2	TOF–TOF	Axima2Analyze, BioMap	Analyze
	AXIMA QIT™	Ion trap TOF	Axima2Analyze, BioMap	Analyze
Waters	MALDI Q-Tof Premier	Qq-oaTOF	MassLynx, BioMap	XML
	MALDI SYNAPT HDMS System	Qq-IMS-oaTOF	MassLynx, BioMap	XML
	MALDI micro MX	TOF	MassLynx, BioMap	XML

Data from all IMS devices can be analyzed with BioMap, through the converter (source: http://www.maldi-msi.org/). The website at http://www.maldi-msi.org/ is a well organized site by Dr. Markus Stoeckli and is recommended for anyone interested in IMS

comparisons and further statistical analysis, co-registration with images of stained sections, etc. This chapter is devoted to introducing the basic use of the BioMap for IMS data analyses.

Originally, BioMap was developed as a software package constructed to visualize and analyze tomographic datasets. The images can be of any origin, including computed tomography (CT), magnetic resonance imaging (MRI), and positron emission tomography (PET). For use also with IMS analysis, it was modified by Dr. Markus Stoeckli [2]. Details of BioMap software are shown in Table 9.2. BioMap was written in interactive data language (IDL); thus, to execute BioMap, it is necessary first to install IDL or IDL VM (IDL Virtual Machine) on one's computer. We install and use IDL VM (without a concern of a licensing fee).

9.2　Required Environment for Installation

The required environment in which we analyze IMS data contains the following components:

1. IDL Virtual Machine (version 6.1 or higher), available at http://www.ittvis.com/idlvm/

Table 9.2 Details of BioMap: image data format and necessary equipment, copyright, and the authors

	Image formats	Analyze 7.5, TIF, DICOM, PNP
Requirements	Windows, Unix, Linux or MAC OS with IDL RT 6.1 or higher	
	Display providing a 24-Bit visual and a resolution of $1{,}280 \times 1{,}024$ or higher	
	256 MB of RAM or more	
Origin, copyright	Novartis	
Author	Martin Rausch, MSI additions by Markus Stoeckli	

Source: "BioMap3x.pdf"; http://www.maldi-msi.org

2. BioMap (as of January 2008, version 3.7.5.5 is the latest version), available at http://www.maldi-msi.org/
3. BioMap Demo Data (sample data: IMS data of whole-rat body section), available at http://www.maldi-msi.org/
4. Data format converter (which can convert your MS data format to "analyze" format)

Note: When the file format of the experimental data is not the "Analyze 7.5" format – which is required by BioMap – it is necessary to convert it into that format, using the data converter software. (When "BioMap Demo Data" is used, this operation is not necessary.) For this purpose, several converter softwares are freely available:

1. Correspondent converter software for Shimadzu Corporation equipment: *AXIMA2Analyze* (available at http://www.maldi-msi.org/) [3]
2. Correspondent converter software for Bruker Daltonics equipment: *Analyze This!* (available at http://www.maldi-msi.org/) [4]

9.3 Setup of BioMap (in Windows XP Operating System)

The procedure to set up BioMap on a Windows XP operating machine is described below. (In Unix, Linux, or Macintosh operating systems, please refer to the "readme.txt and BioMap3x.pdf" of the downloaded BioMap-related file) The procedure of the setup shown here is an example, and it is possible to change according to the file name and the layered structures according to preferences.

1. Double-click the installation file of IDL VM and then the installation begins. (Default settings are recommended for the installation destination.)
2. Unzip the downloaded "zipped BioMap" file.
3. Create a **"C:\Biomap"** folder, and move all unzipped files to this folder.
4. Next, move all files that exist in the "C:\Biomap\Setting" folder to the "C:\ Documents and Settings**username**\Application Data\Biomap" folder.

Note: The "username" changes with each individual.

5. A shortcut file of the "Biomap.sav" in "C:\Biomap\bin" folder is produced, and places the shortcut file on the computer's desktop.

6. The window of IDL VM appears when the "Shortcut to Biomap.sav" file is double-clicked and the program is started up; click "Click to continue" to start.

7. When BioMap is started for the first time, a download ID is displayed in the upper part of the window. At this point, all functions are locked. Then, it is necessary to register at http://www.maldi-msi.org to activate them (Fig. 9.1).

8. When registration at the web site is completed, the "licenserc." file will be sent to the registered e-mail address. Move the "licenserc." File to the "C:\Biomap" folder.

9. Restart the BioMap.

10. If the software is activated by the user's registration, the upper part of the window where the ID was displayed becomes "BioMap 3.7.5.5 provided by Novartis Institutes for BioMedical Research." The BioMap software is now available for analysis (Fig. 9.2).

11. Confirm the current specification of "main path" of BioMap (normally, the main path is placed in C:\Biomap\). In "Edit→Defaults→Folders" (menu bar of BioMap), the present setting of the main path of BioMap can be confirmed. Click the "Folders" of tag window in the "Edit default" window, and enter the correct main path; see Fig. 9.3.

Note: If the main path of BioMap is not correctly specified, an alert "The BioMap main path may not be set up correctly, you must set the variable BioMapDir in your uparsrc file manually" appears, and part of the function is not available.

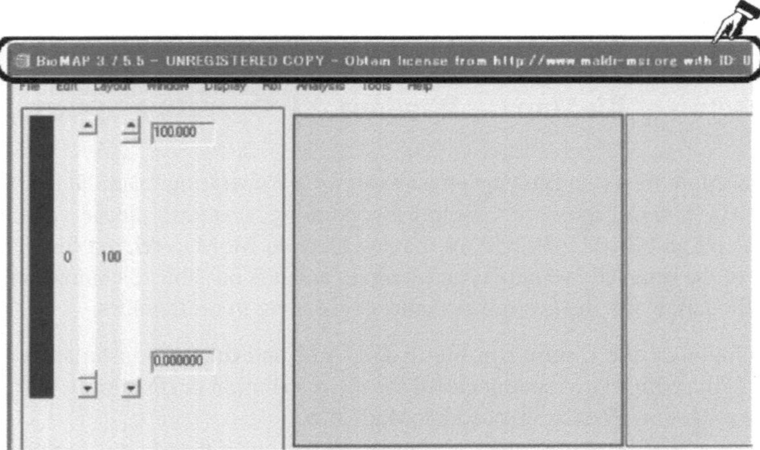

Fig. 9.1 BioMap is functionally locked before user registration. When "ID" is displayed in the upper part of the window, the functions of BioMap cannot be used. Activating the software at http://www.maldi-msi.org is necessary

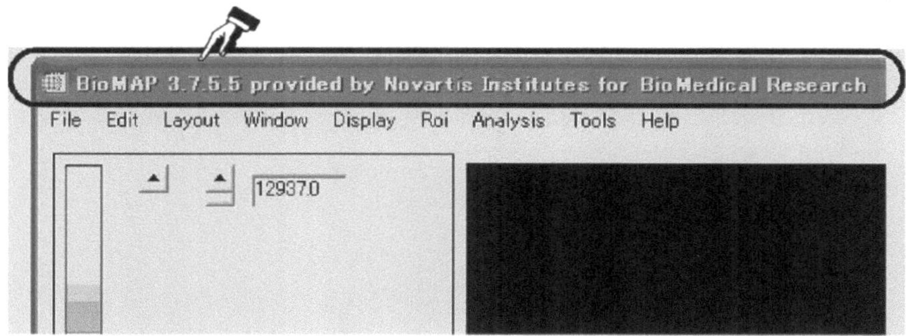

Fig. 9.2 The locked functions of BioMap are released. When "BioMap 3.7.5.5 provided by Novartis Institutes for BioMedical Research" is displayed in the upper part of the window, one knows that the activation has been successful

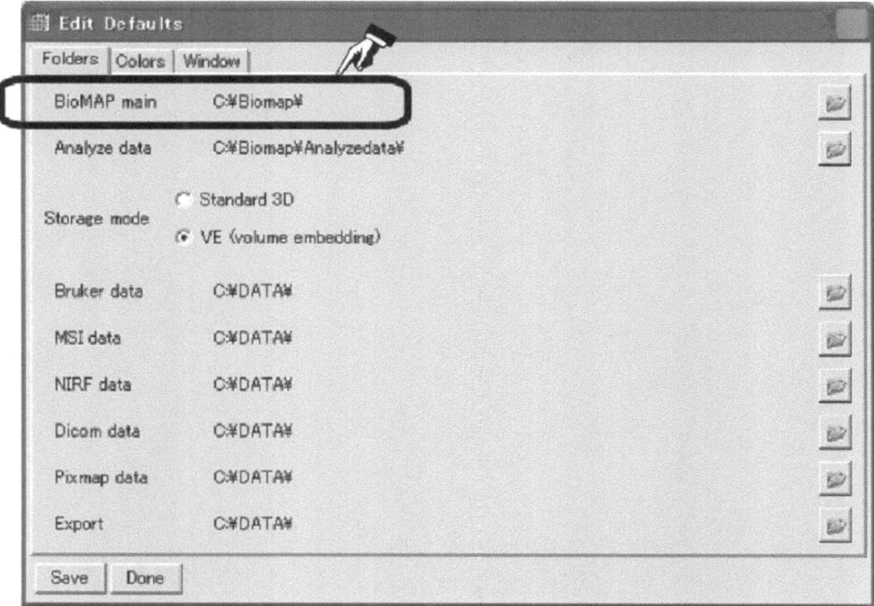

Fig. 9.3 Example of "BioMap main path" setting

9.4   Example of Ion Image Reconstruction from IMS Data

In this section, we show how to reconstruct a two-dimensional ion distribution image from BioMap Demo Data (sample data: IMS data of whole-rat body section). Figure 9.4 shows examples of data presentations. The abundance of ion is shown as a gradation of brightness (white and black represent high and low

                                                                            N. Hosokawa et al.

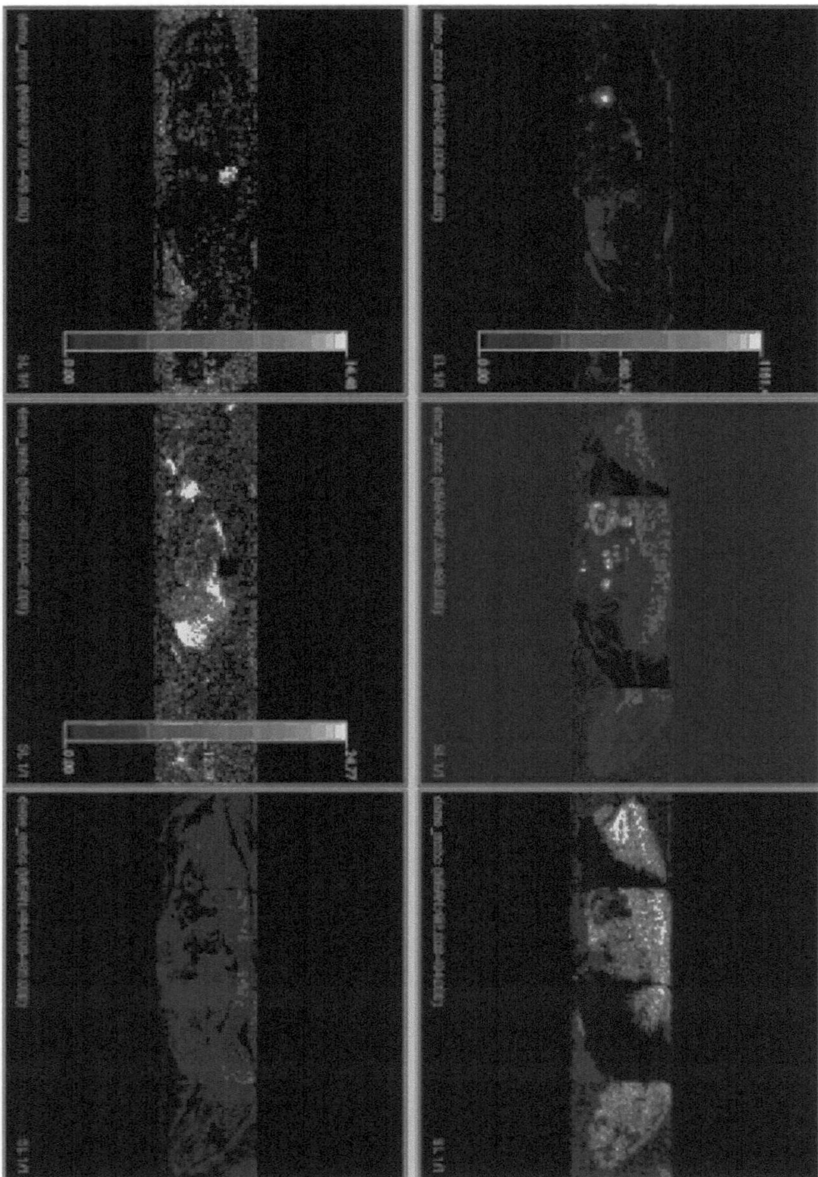

Fig. 9.4 Example of two-dimensional (2D) strength distribution of arbitrary *m/z* in whole-rat body section. It is sequentially displayed as *m/z* 497, *m/z* 488, *m/z* 493, *m/z* 498, *m/z* 481, and *m/z* 485, from the *left upper* to *right lower* corner

intensity, respectively) corresponding to their peak intensities. As can be seen, the intensity scale is defined for each ion distribution image, i.e., the same brightness does not mean the same intensity value among images.

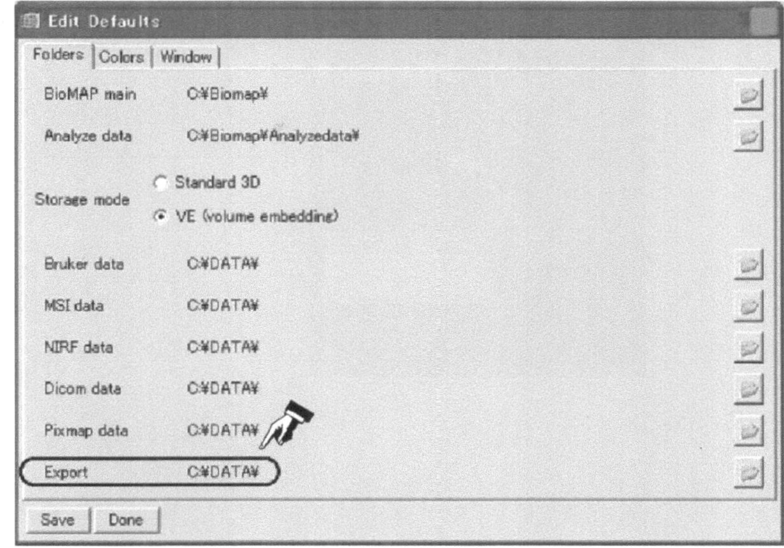

Fig. 9.5 Example of setting export (BioMap main path)

1. In "Edit→Defaults→Folders" (menu bar of BioMap), specify the folder for the input/output (Fig. 9.5).

   Note: If incorrect specification, for example, if the folder of C:\DATA does not exist, users cannot export function of any image or spectra.

2. In "File→Import→MSI" (menu bar of BioMap), select the "demo.img" file in the BioMap Demo Data folder. It is necessary to adjust the m/z range to load, bin size, etc., to achieve a suitable Import Data Size to your computer.

   Note: When the "Import Data Size" exceeds 300 MB, the software does not work at our system (which is Windows XP, Pentium 4 Dual CPU 3.00 GHz and 3.01 GHz, RAM 1.00 GB).

3. When "Analysis→Plot→Point" (menu bar of BioMap) is opened, a line cross appears in "window 1," and spectrum information corresponding to the indicated tissue location is displayed in the "Plot Tool" window. By moving the intersection, the mass spectrum at the indicated pixel can be seen.

4. It is also possible to display the average of a spectrum group within the square area specified by right-clicking with the mouse. By averaging the spectra of several data points, researchers can find their mass peak of interest from a higher-quality spectrum.

   In Fig. 9.6, the averaged spectrum from the square area (white frame: five pixels) in window 1 is displayed in the "Plot Tool" window.

5. To display the ion distribution image of interest, perform the following operation with the mouse within the "Plot Tool" window:

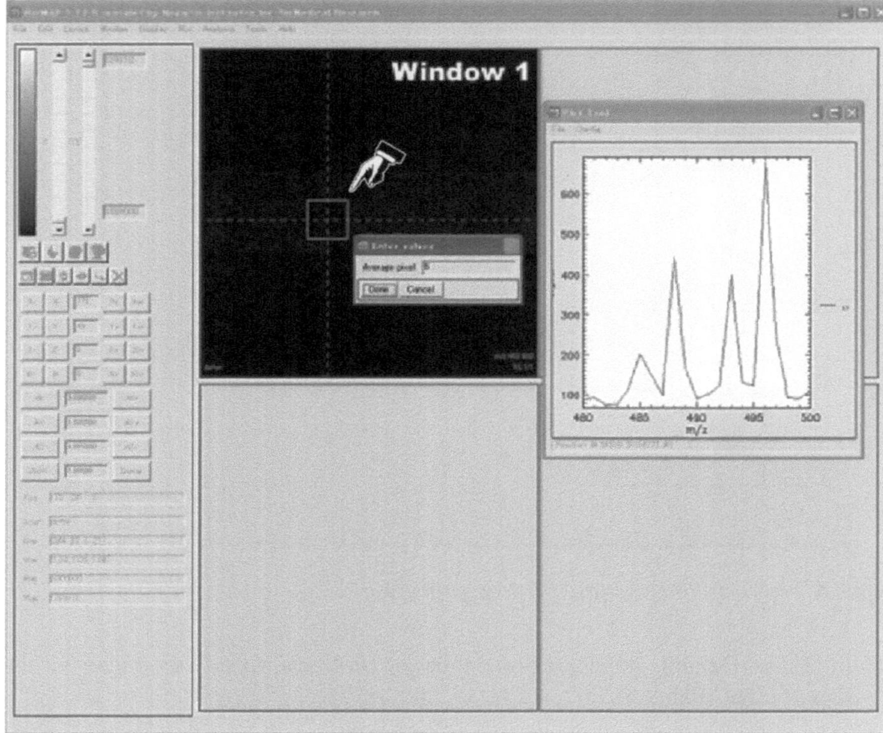

Fig. 9.6 Example of average display of all spectra in white frame

6. Expand the mass spectrum to select a single mass peak, and the selected icon is displayed as a two-dimensional ion distribution image in (Fig. 9.7).
   To do this operation, use three mouse buttons:
   (a) *Mouse left drag:* Used to quickly display the distribution image of the selected m/z value.
   (b) *Mouse right drag:* Used to expand the mass spectrum. The selected part (shown in red) is expanded and displayed. It returns to the overall view by double-clicking it.
   (c) *Mouse wheel button:* Used to select mass peak (i.e., m/z window) to image. The distribution image of the selected m/z window (shown in green) (i.e., the highest ion intensity in the window, at each pixel) is displayed on specified screen.

   Note: Users can select another command to display, such as the value of the integral in the m/z window.

7. Figure 9.8 shows an example of an ion distribution image at m/z 488 from the BioMap Demo Data. User can display the selected ion distribution image in the "Plot tool" window, on a specified screen (in this case, window 2).

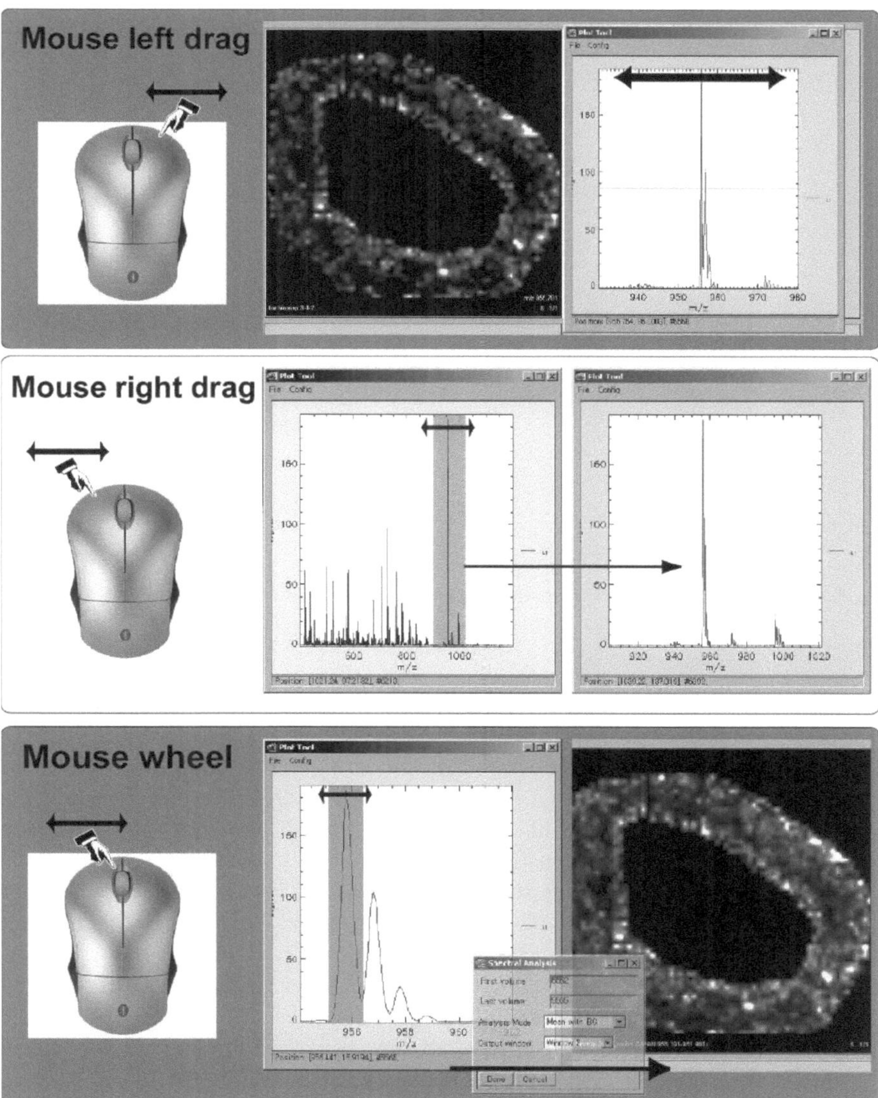

Fig. 9.7 The use of mouse for ion image generation (see main text for details)

8. To process the created image, the user should make "window 2" the work screen (i.e., active screen). To do this, close the "Plot Tool" window and click the left mouse button on window 2. When window 2 becomes a work screen, its outer border is framed in red.

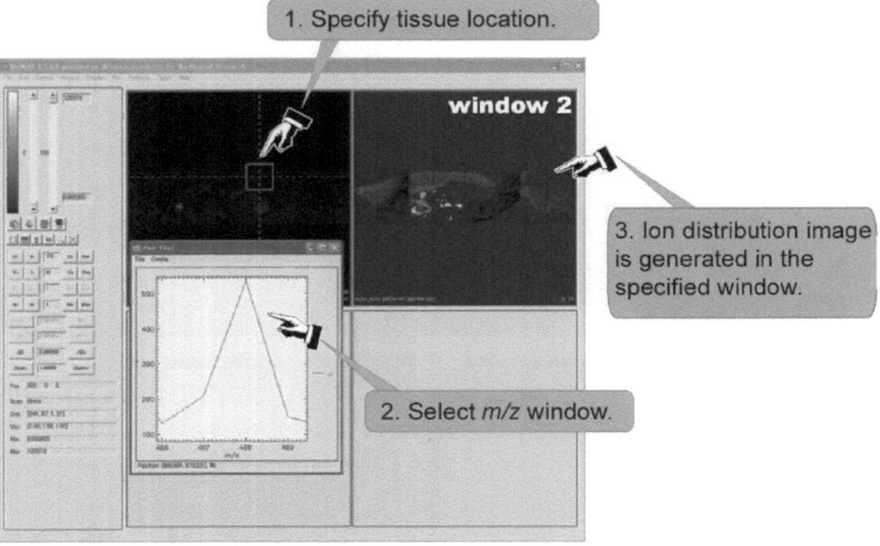

Fig. 9.8 Two-dimensional ion distribution image of *m/z* 488, shown in window 2

9. If the generated ion image is not clear, the user can process the image by adjustment of image contrast and color intensity scale.
10. To adjust image contrast, user can adjust the minimum intensity and full intensity threshold of a result filter (Fig. 9.9a). With the left slider, you can adjust the minimum intensity threshold between 0% and 100%. With the right slider, you can also adjust the full intensity threshold.
11. To change the gray intensity scale to the color scale, just left-click the scale bar or select "Window Color" table in the menu bar of BioMap, and then select your favorite type of scale color (Fig. 9.9b).
12. In "File Export Image" (menu bar of BioMap), an result images can be exported in a specified file path as a picture-format file.
13. In "Layout" (menu bar of BioMap), the number of display images is adjusted and a more effective image display becomes possible (Fig. 9.10).

9.5 Comparison Analysis of Mass Spectra in Each ROI

BioMap allows the differential data analysis of the mass spectra between ROIs. After two or more ROIs have been specified in a tissue section, the average mass spectrum of each ROI can be displayed and a comparison can be performed, i.e., the characteristic mass peak for specific internal organs, cells or any interest region can be identified. The compared mass-spectral data can be exported into text

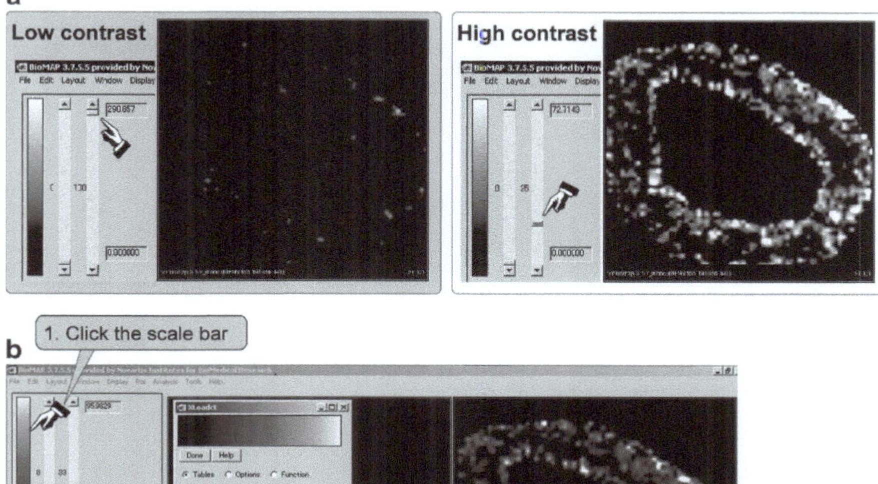

Fig. 9.9 Experimenters can adjust the image contrast (**a**) and change the type of scale bar (**b**)

format, and thus users can further analyze them by other statistical software, such as Excel and Origin software.

The procedure for differential analysis can be performed as follows:

1. In "File→Import→MSI" (menu bar of BicMap), select the "demo.img" file from the BioMap Demo Data folder (or your original data file).
2. If you have an optical microscopic image (such as a dye-stained tissue image), activate another screen by right-clicking, and the optical image for the reference can be imported in "File→Import→Pixmap" (menu bar of BioMap).
3. In this case, the peak of *m/z* 488 is selected and the distribution image is displayed in window 2. Next, select the window 2 as the work screen.
4. In "Window→Zoom" (menu bar of BioMap), it is possible to magnify the image to define small ROIs.
5. By select "Roi→Draw" table (menu bar of BioMap), the "Draw ROI Tool" window appears. Them define the ROIs on the image, and click "Done."
6. In the table of "Roi→Name" (menu bar of BioMap), it is also possible to name the area. In "Roi→Remove" (menu bar of BioMap), it is also possible to delete the area previously defined.

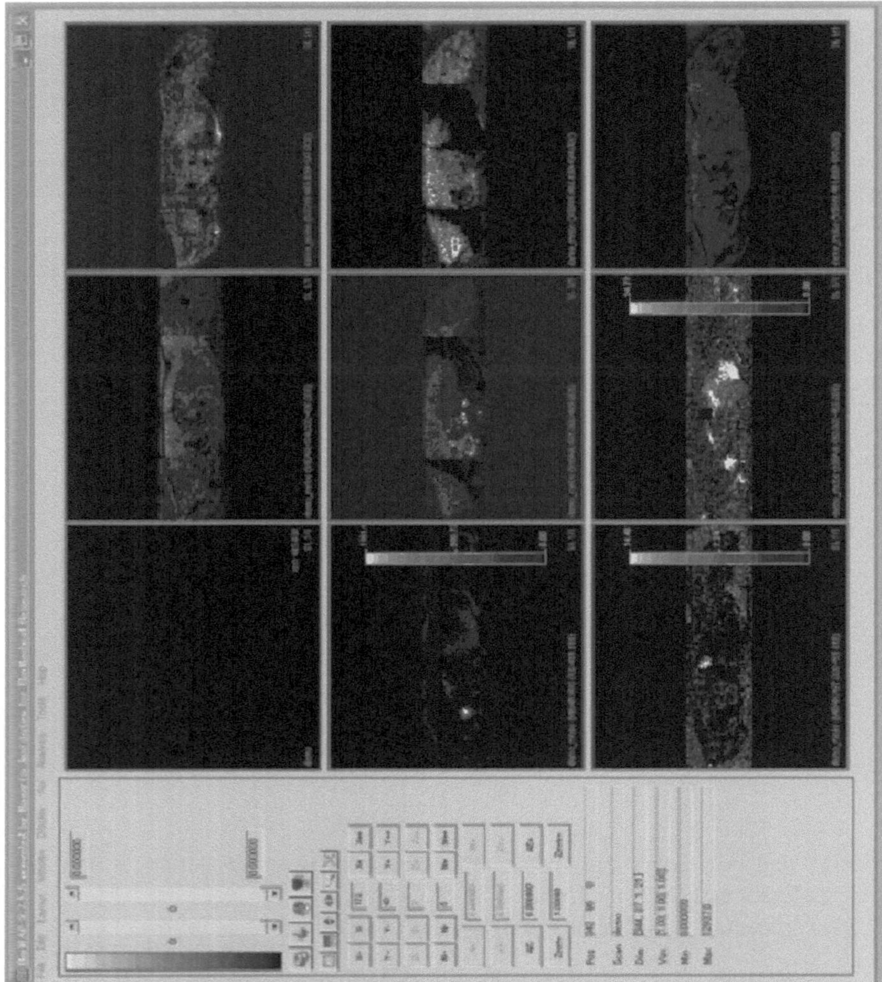

Fig. 9.10 Example of display layout with 3×3. The number of windows can be increased from 1×1 to 5×5

7. To analyze the mass spectra from each ROI, the ROIs should be copied on the window 1. By select of "Edit→Copy" command, defined ROIs can be copied. Make the window 1 as a work screen, and paste the ROIs by "Edit→Paste→Roi" (menu bar of BioMap) (Fig. 9.11).
8. By selection of "Analysis→Plot→Roi" (menu bar of BioMap) command, the averaged spectrum from each ROI is displayed (Fig. 9.12). By comparison of mass spectra, the unique mass peak in each ROI can be easily found.
9. Via "File→Export" (in the "Plot Tool" window), the image data and spectrum data of each area can be stored on the computer as a text format. This operation

makes it possible to perform a quantitative comparison between spectrum groups with other statistical software (e.g., Microsoft Excel and Origin).

10. Proceed creation of the ion distribution image of identified mass peaks and confirm the uniqueness.

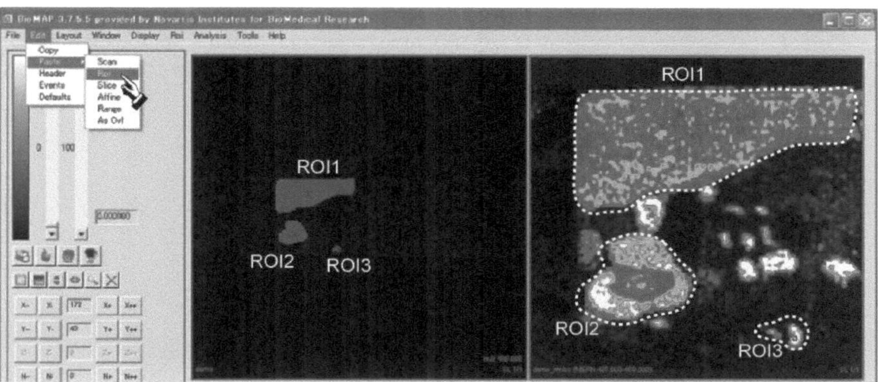

Fig. 9.11 To analyze the mass spectra from each ROI, the ROIs should be copied on window 1. In this figure, ROIs are defined on window 2 in which magnified ion image at *m/z* 488 is displayed, and then they are copied on window 1

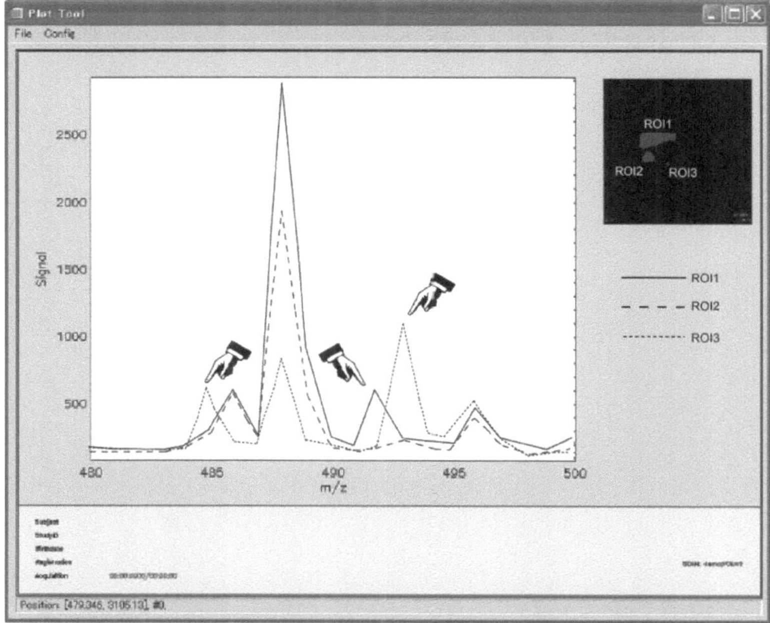

Fig. 9.12 A particular peak is indicated by an *arrow*

References

1. Stoeckli M. From http://www.maldi-msi.org/
2. Stoeckli M, Staab D, Staufenbiel M, et al. (2002) Molecular imaging of amyloid beta peptides in mouse brain sections using mass spectrometry. Anal Biochem 311:33–39
3. Shimma S, Sugiura Y, Hayasaka T, et al. (2008) Mass imaging and identification of biomolecules with MALDI-QIT-TOF-based system. Anal Chem 80:878–885
4. Clerens S, Ceuppens R, Arckens L (2006) CreateTarget and Analyze This!: new software assisting imaging mass spectrometry on Bruker Reflex IV and Ultraflex II instruments. Rapid Commun Mass Spectrom 20:3061–3066

Chapter 10
Statistical Procedure for IMS Data Analysis

Yuki Sugiura and Mitsutoshi Setou

Abstract In MALDI-IMS of tissue samples, since the tissue contains an enormous variety of biomolecules, a complex mass spectrum with hundreds of to a thousand peaks can be obtained from a single data point. Furthermore, several thousands of spectra with spatial data are obtained at one IMS experiment. Because of the complexity and enormousness of the IMS dataset, manual processing of the dataset to obtain significant information (e.g., identification of disease-specific mass signature) is not a realistic procedure. In this regard, today, multivariate analysis becomes a powerful tool in IMS data analysis. In this chapter, we describe an unsupervised multivariate data analysis technique that enables us to sort the data sets without any reference information. Particularly, two major methods that are related to IMS, namely, hierarchical clustering and principal component analysis (PCA), are described in detail with examples. Finally a basic procedure for PCA with familiar software (such as Microsoft Excel) is introduced.

10.1 Introduction

A single mass spectrum contains a lot of useful information. In an imaging mass spectrometry (IMS), the spectra are further added to spatial information by scanning the tissue sample two-dimensionally. Moreover, a typical IMS-mass spectrum contains hundreds to a thousand peaks because tissue sections contain an enormous number of biomolecules. Furthermore, the number of obtained spectra could be several thousands in one IMS measurement. The recent advances of IMS regarding

Y. Sugiura

Department of Bioscience and Biotechnology, Tokyo Institute of Technology,
4259 Nagatsuta-cho, Midori-ku, Yokohama, Kanagawa 226-8501, Japan

Y. Sugiura and M. Setou (✉)
Department of Molecular Anatomy, Hamamatsu University School of Medicine,
1-20-1 Handayama, Higashi-ku, Hamamatsu, Shizuoka 431-3192, Japan
e-mail: setou@hama-med.ac.jp

M. Setou (ed.), *Imaging Mass Spectrometry: Protocols for Mass Microscopy*,
DOI 10.1007/978-4-431-09425-8_10, © Springer 2010

simultaneous detection of numerous molecules (i.e., large number of mass peaks) at high spatial resolution (i.e., large number of data points) further enlarge the volume of IMS datasets. Because of such enormity of the IMS dataset, the development of statistical analyses is essential, and several multivariate analyses have been developed as useful tools for extracting important information. In particular, to date, "unsupervised" multivariate data analysis techniques have been widely utilized; they enable one to sort datasets without any reference information [1–4].

As Yanagisawa et al. reported in an early study [4], unsupervised multivariate analysis – particularly hierarchical clustering – helps one extract clinically important information for diagnosis/prognosis purposes, from a large-scale dataset. Also, principal component analysis (PCA) [1–3] and independent component analysis (ICA) [5] have been used to reduce data, thus enabling the extraction of the specific mass peaks [6, 7] and tissue locations of interest (e.g., normal tissue region vs. cancerous region) [7]. More recently, a more readily interpreted biological/clinical method using probabilistic latent semantic analysis (pLSA) has been developed [8].

In this chapter, we describe a quite basic protocol for an unsupervised multivariate data analysis technique, particularly PCA, using only Microsoft Excel and free software. Although several useful software packages commercially available provide much fast and easier data analysis with automatic calculation (just click "Run" command; see Chap. 11), the aim of this chapter is introduction of statistical analysis with minimal procedure for one who is not statistically expert. It is strongly recommended that each experimenter perform the analysis himself or herself using standard statistical tools such as Microsoft Excel; such experience should help the experimenters in later performing more complicated analyses governed by software packages that proceed automatically with calculation.

10.2   IMS Linked to Multivariate Analysis

Multivariate analysis is applied to a dataset that involves more than one statistical variable at a time. Because large numbers of variables are involved in matrix-assisted laser desorption/ionization (MALDI)-IMS – as a result of the mass spectrum combined with spatial information, researchers find multivariate analysis effective in extracting biological/clinical information. Hierarchical clustering is used to sort datasets into several clusters, according to similarities among variables. The traditional representation of this hierarchy is a tree (called a dendrogram), with individual elements at one end and a single cluster containing every element at the other. Agglomerative algorithms begin at the leaves of the tree, whereas divisive algorithms begin at the root [9].

In IMS (or direct tissue profiling with MS), each of the MS profiles could be clustered, based on the m/z value and its intensity, i.e., peak expression pattern. Employing this technique, one can identify the mass peaks that are different between control and diseased samples (or regions), thus possibly identifying biomarkers for specific patients by the molecular signature. Yanagisawa et al. performed direct MS analysis of lung tissue obtained from non-small cell lung cancer (NSCLC) or nontumor patients and processed the direct tissue profiling data via the clustering [4]. As shown in Fig. 10.1 (top panel), the profiles of NSCLC and nontumor patients can be clearly distinguished. In addition, the primary alteration can be discriminated from others (bottom panel). Such an unsupervised clustering analysis enables not only sorting data into clusters but also extracting reliable markers according to the statistical criteria. In the technical aspect, in this example, by converting mass spectra into the data matrix of a peak intensity list, the authors reduced the volume of data, enabling an efficient analysis.

PCA, on the other hand, has been utilized to understand the overview of the spatial molecular distribution patterns [1–3]. We do not describe the detailed mathematical theory for reasons of space limitation, but in brief, it is a statistical method that merges the data containing multiple elements into low-dimensional data.

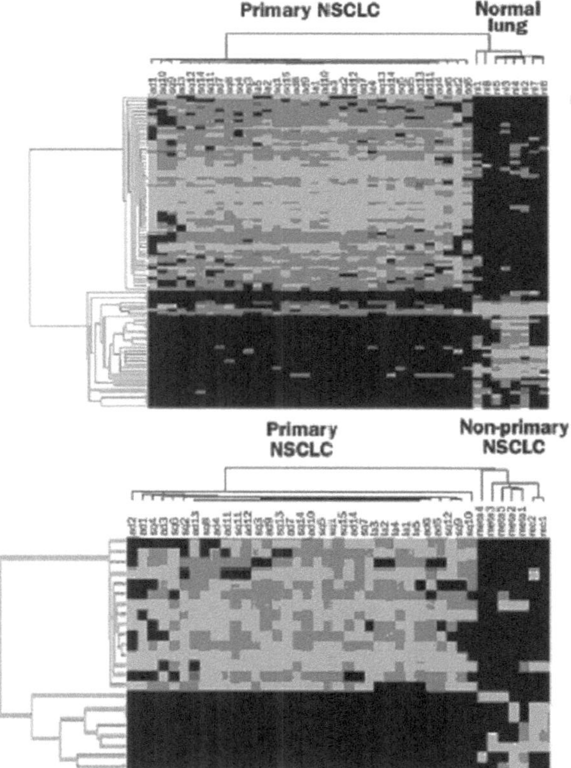

Fig. 10.1 An example of clustering according to peak intensity, calculated from mass spectra. Thirty-four non-small cell lung cancer (*NSCLC*) patients were clearly distinguished from eight controls. In addition, primary patients could be distinguished from nonprimary patients [4] (Reprinted from Yanagisawa et al., Lancet 362:433–439.)

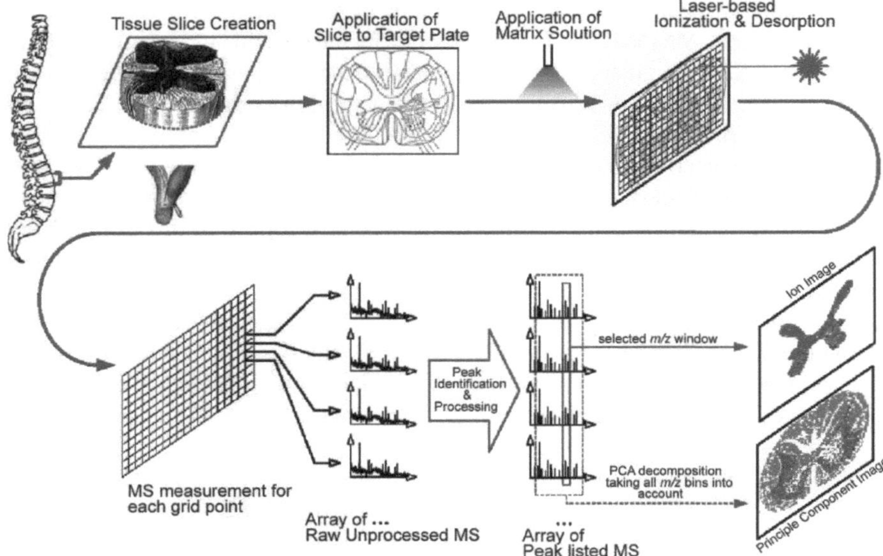

Fig. 10.2 After performing standard IMS using spinal cord sections, each peak was detected to obtain two-dimensional (2D) images of ions. At the same time, PCA analyses were performed by using intensity values derived from all the mass peaks [3] (Reprinted from Plas et al., 2007 IEEE/NIH Life Science Systems and Applications Workshop, pp 209–212.)

It reduces a large set of variables to a small set of variables called "principal factors," which are linear combinations of the original variables.

Figure 10.2 shows an example of PCA-coupled IMS data analysis. In the strategy reported by Plas et al., spectra obtained by IMS are processed to peak detection, and based on the generated peak list, two-dimensional (2D) ion images were reconstructed and PCA decomposition was performed. PCA images (i.e., 2D ion intensity map of principal component score on the tissue section) were utilized to find trends of proteomic distribution patterns. By utilizing IMS-PCA on spinal cord tissue section, Plas et al. demonstrated that proteomic composition in the "butterfly-shaped" posterior column of the spinal cord differs considerably from that in other regions of the spinal cord, according to statistical criteria [3].

Figure 10.3 shows another example, in which the IMS-PCA of a breast cancer section was performed. In this case, PCA revealed that the largest spectral differences (i.e., the largest difference in proteomic composition) were observed between connective tissue and cancer area (in the principal component 1 image, panel e). Furthermore, the second largest differences in protein expression pattern were observed within two tumor cell populations, which are HER2-positive/negative cells (panel b), revealed by the principal component 2 image (panel f) [10].

PCA is also helpful in identifying the meaning variables, for example, identifying which mass peaks are responsible for making difference between normal and drug-treated tissue regions. Such mass peaks can be candidates for the bio-

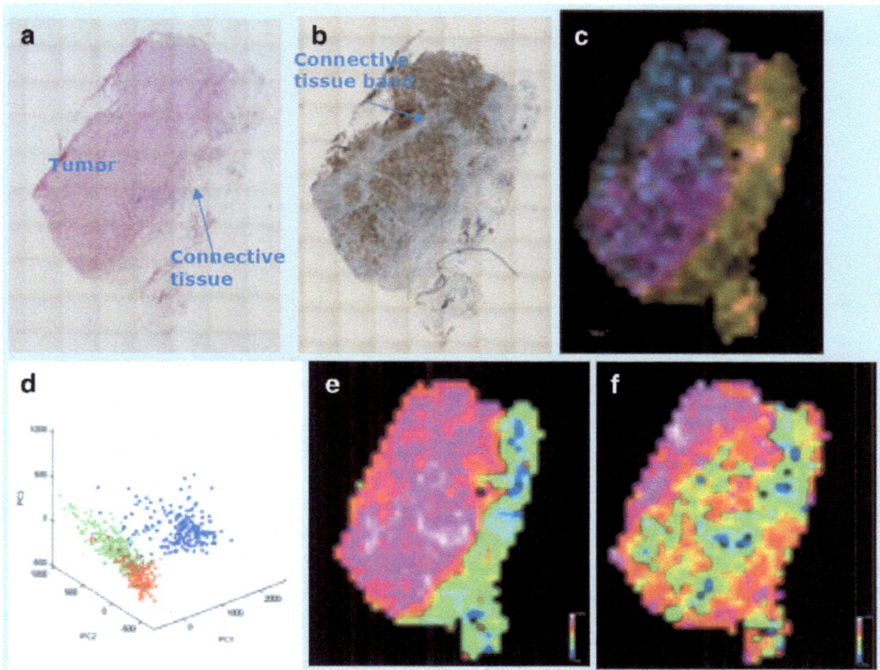

Fig. 10.3 IMS linked PCA with breast cancer section. Optical images of hematoxylin and eosin (H&E)-stained (**a**) and HER2-immunostained (**b**) breast cancer sections. **c** Ion image at several mass peaks were stained with *blue* and *red*. **d** Graph in which principal component scores for PC1, PC2, and PC3 are plotted. **e** Principal component 1 image. According to the value of principal component score calculated for each spectrum at each tissue location, pixels were stained with pseudo-color. **f** Principal component 2 image (Kindly permitted by PD Dr. med. Axel Walch, Institute for Pathology, Helmholtz Center Munich, German Research Center for Environmental Health (GmbH) Ingolstadter Landstrasse 1, 85764 Oberschleissheim, Germany)

marker [11, 12]. Prideaux et al. showed the altered metabolic profiling of porcine skin after treatment with a commercial 0.1% hydrocortisone cream. By utilizing PCA, they successfully sorted the two groups of skin MS profiles with/without drug treatment and identified the altered peak expressions (Fig. 10.4) [6]. Such direct MS profile coupling PCA can be applied to analysis in micro-tissue domains as small as a laser spot (diameter >100 μm) that are difficult to separate and analyze by using conventional approaches [6, 7].

10.3 IMS-PCA on the Genetically Manipulated Mouse Brain

In the following sections, we introduce a simple protocol to identify mass peaks of which difference between samples by PCA. We used only Microsoft Excel and SpecAlign software, the latter of which is a graphic computational tool for

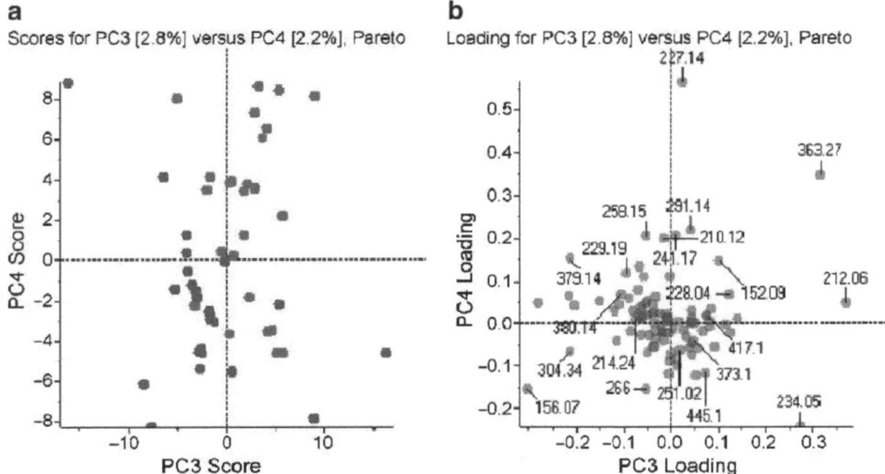

Fig. 10.4 **a** Principal component analysis of 25 spectra taken from inside and 25 spectra from outside the area of porcine skin treated with a commercial 0.1% hydrocortisone cream. In the score plot, a clear grouping of treated (*red spots*) and untreated (*blue squares*) spectra is observed. **b** The factor loading plot indicates that m/z 363.28, the $[M + H]^+$ ion for hydrocortisone, is a major contributor to differentiation among the groups (Reprinted from Prideaux et al., Int J Mass Spectrom 260:243–251.)

spectrum processing. (SpecAlign is free of charge and available for download from Wong et al., from http://ptcl.chem.ox.ac.uk/jwong/specalign [13].)

Here, we analyzed animals with distinct genetic backgrounds; they were either wild type (WT) or SCRAPPER (Scr) knockout (KO) mice lacking a gene coding for a ubiquitin ligase [14]. We applied IMS-linked PCA, to compare the proteomic composition of WT and Scr-KO mice brains and further searched for substances differentiating the two genotypes.

10.4 IMS of WT and Scr-KO Mouse Brain Sections

In Scr-KO mice, obvious pathological features were observed, particularly in their brain. For example (Fig. 10.5), hematoxylin and eosin (H&E) staining of Scr-KO brain tissue revealed the existence of sponge-like degeneration. It has been reported that the mutant mice of other ubiquitin E3 ligase exhibit an age-dependent neuropathology, including spongiform degeneration [15]. The release of neurotransmitter in Scr-KO mice is abnormal [14]; such neural dysfunction possibly induced the histological degeneration observed in Fig. 10.5.

There are several ways to determine the molecules that are involved in such histological degeneration. For example, in immunohistochemistry and immunoblotting, specific alterations of proteins with high sensitivity and specificity can be revealed; however, analyses require specific antibodies to the candidate protein.

corpus striatum

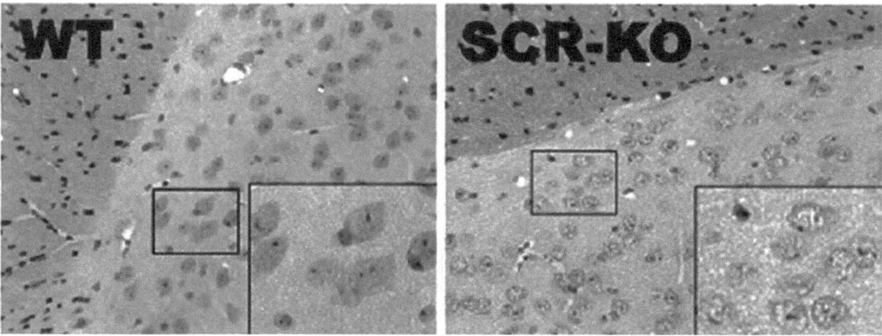

Fig. 10.5 Optical image of hematoxylin and eosin (H&E)-stained corpus striatum region of wild-type (*WT*) and Scr-knockout (*SCR-KO*) mice [7] (Reprinted from Yao et al., Proteomics 8:3692–3701.)

Because generation of the antibodies requires large cost and much time, one must determine the small number of candidate molecules that might be involved in the defect before the experiment.

In this regard, MS is able to detect molecular alterations without any tags or probes, even of unexpected molecules. In this context, MS-based imaging can identify such differences with spatial information, even within micro-tissue regions that are difficult to dissect out.

10.4.1   Materials Used and Measurement Conditions

The following are the details of an IMS experiment using sagittal brain sections obtained from WT and Scr-KO mice:

- Brain sections, sliced at 5-μm thickness [16], mounted on ITO glass slide (Bruker Daltonics)
- 70% Ethanol for rinsing tissue sections, 30 s × two times [17]
- Matrix: Sinapic acid (Bruker Daltonics) [25 mg ml^{-1} in 0.1% trifluoroacetic acid (TFA), 50% acetonitrile (v/v)], applied to section by spray-coating method
- Mass spectrometer: MALDI-TOF/TOF-type instrument (Ultra Flex 2, Bruker Daltonics)
- Measurement modes: positive-ion detection mode, linear mode
- Laser interval: 80 μm
- Software used for image reconstitution and spectrum-extraction from region of interest: flexImaging 2.0 (Bruker Daltonics)

Acquired IMS data were sorted and preprocessed before PCA, as follows:

- Following IMS measurement, mass spectra were collected from each brain region of WT and Scr-KO mice (using flexImaging 2.0 software); in this study, those regions were composed of the cerebral cortex, pons, hypothalamus, and corpus striatum (see Fig. 10.13 later in this chapter).

- The spectral data were then formatted into ASCII code, using flexAnalysis 3.0 software, to allow processing by SpecAlign software.
- Next, the converted spectra were normalized to equalize total ion current by using SpecAlign software [13]. This is an important process, as it rescales the sample-to-sample variability of the peak intensity values before proceeding to statistical analysis [6, 18, 19].

In data analysis procedures without statistical methods, we usually averaged the spectra of each region and visually compared the mass peaks between samples one by one. With such visual comparisons of spectra (Fig. 10.6), we were certainly able to find differences among the peak expressions, as indicated by the arrows; this, in turn, indicates the abnormal expression/suppression of proteins in KO mice. However, such methodology is inefficient, especially when one is analyzing a large number of regions and/or many tissue sections.

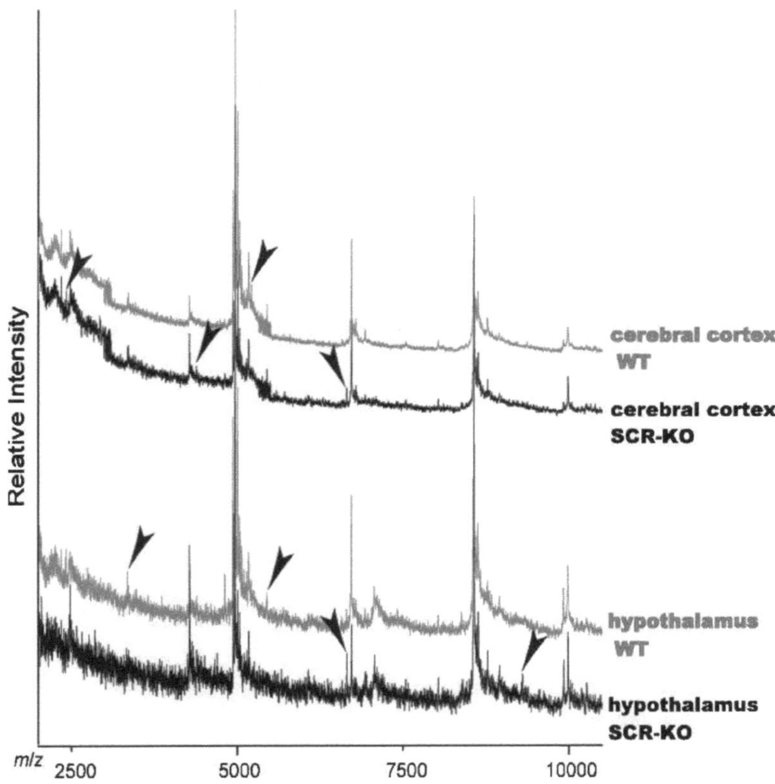

Fig. 10.6 Comparison of normalized spectra following extractions from *WT* and *SCR-KO* mice [7] (Reprinted from Yao et al., Proteomics 8:3692–3701.)

10.4.2 IMS-PCA of WT and Scr-KO Mouse Brains

Below, we describe IMS-linked PCA to compare the proteomic composition of the corpus striatum region, in which severe pathological alterations were found in Scr-KO mice (see Fig. 10.5).

For PCA, to reduce the amount of calculation required, we generated mass peak lists from the sets of extracted spectra rather than use of every continuous m/z value; For this operation, peak detection function of SpecAlign software was used [13]. Although such a scenario would be ideal, using every m/z value of every spectrum as a variable would claim too many calculations. Figure 10.7 shows the peak list file used for PCA, created by importing peak list files into Microsoft Excel [arranged as x-axis$=m/z$ (validate), y-axis$=$spectrum (case)].

Next, PCA was performed by Microsoft Excel add-in software. Several free or low-cost software packages that can be used as add-in or macro tools for Microsoft Excel are available. Generally, we must define the "case" and "variable" for PCA calculation; in this case, we defined each mass spectrum with spatial information as the case, and each includes approximately 80 distinct mass peak intensities (variables).

10.4.3 Data Interpretation of PCA

10.4.3.1 Interpretation of Component Scores

As a result of undertaking PCA, several parameters can be obtained. Among those, the *component score* and *factor loading* are important postanalysis steps. A component score is calculated for each mass spectrum; all are defined for each principal component (e.g., for PC1, PC2....) (Fig. 10.8).

Those component scores can be plotted two-dimensionally to facilitate interpretation of the PCA results. In Figs. 10.9 and 10.10, component scores for PC1 and PC2 are plotted on the x-axis and y-axis, respectively. and each dot represents a spectrum from a distinct tissue sample or location. What is important to note is whether two populations of spectra ($=$dot) obtained from distinct samples (e.g., normal vs. diseased) are "spatially separated" on the graph. If they are separated (see Fig. 10.9a), it means that the molecular expression patterns of these two regions were statistically different from each other. If not, PCA failed to extract the statistical differences between the two populations (see Fig. 10.9b).

In Fig. 10.10, open and filled spots indicate spectra obtained from WT and Scr-KO mice, respectively. Notably, two populations of spots are spatially separated on this graph (as represented by open and filled ellipses), indicating that those two sets of spectra from WT and Scr-KO mice tissue contain proteomic differences. We focused on the y-axis (PC2) that separates nicely between the two populations: one could find that the spectra of WT mice had positive component scores against PC2 whereas Scr-KO mice had negative scores.

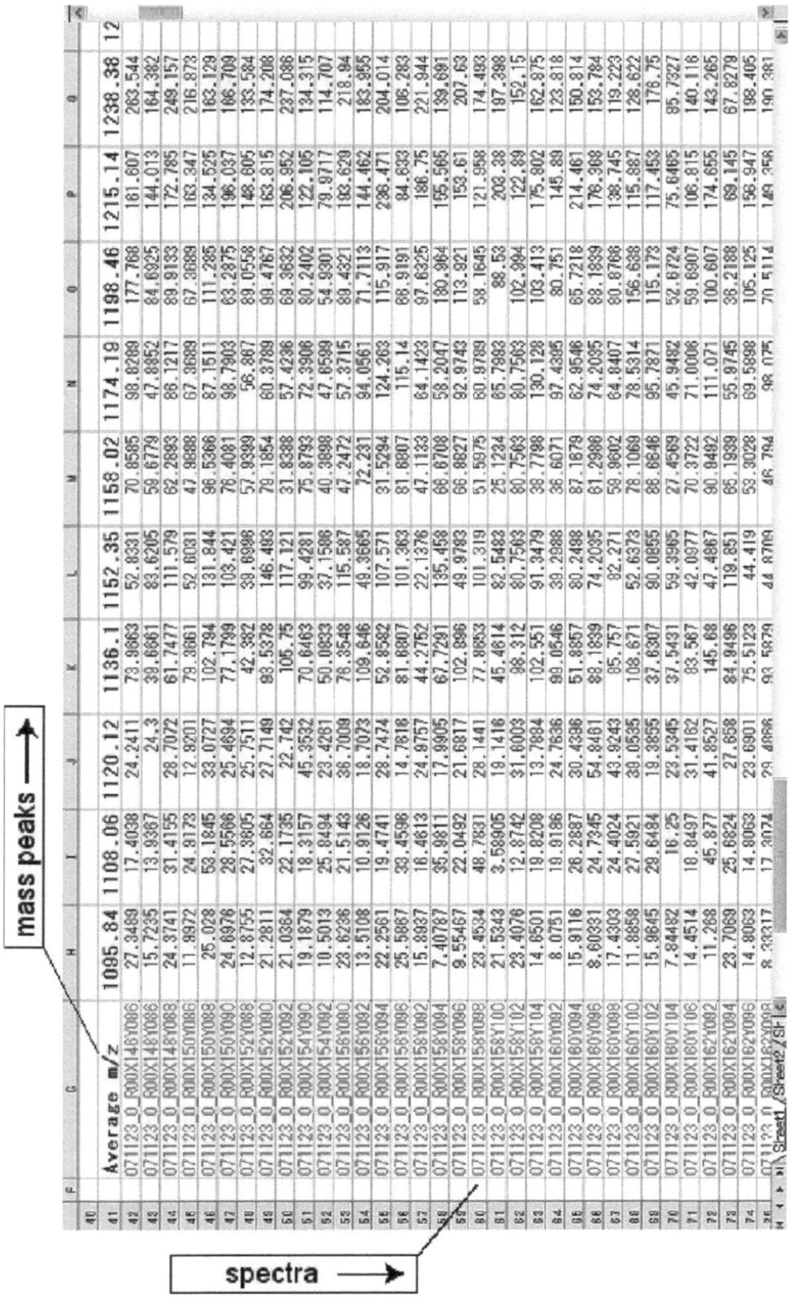

mass peaks →

spectra →

Average m/z	1095.84	1108.06	1120.12	1136.1	1152.35	1158.02	1174.19	1198.46	1215.14	1238.38	12
071123_0_R00X146Y086	27.3489	17.4038	24.2411	73.9663	52.8331	70.8585	98.8289	177.768	161.607	263.544	
071123_0_R00X148Y086	15.7235	13.9367	24.3	39.6661	83.6205	59.6779	47.8852	84.6925	144.013	164.382	
071123_0_R00X148Y088	24.3741	31.4155	28.7072	61.7477	111.579	62.2893	86.1217	89.9133	172.785	249.157	
071123_0_R00X150Y086	11.8972	24.9173	12.9201	79.3661	52.6031	47.9898	67.3689	67.3689	163.347	216.873	
071123_0_R00X150Y088	25.028	53.1845	33.0727	102.794	131.844	96.5366	87.1511	111.285	134.525	163.129	
071123_0_R00X150Y090	24.6976	28.5566	25.4694	77.1799	103.421	76.4081	98.7903	63.2875	196.037	166.709	
071123_0_R00X152Y088	12.8755	27.3605	25.7511	42.382	39.6996	57.9399	56.867	89.0558	148.605	133.584	
071123_0_R00X152Y090	21.2811	32.664	27.7149	93.5378	146.493	79.1854	60.3789	99.4767	163.815	174.208	
071123_0_R00X152Y092	21.0364	22.1735	22.742	105.75	117.121	31.8388	57.4236	69.3632	206.952	237.086	
071123_0_R00X154Y090	19.1879	18.3157	45.3532	70.6463	99.4281	75.8793	72.3906	80.2402	122.105	134.315	
071123_0_R00X154Y092	10.5013	25.8494	23.4261	50.0833	37.1586	40.3898	47.6599	54.9301	79.9717	114.707	
071123_0_R00X156Y090	23.6236	21.5143	36.7009	76.3548	115.587	47.2472	57.3715	89.4321	183.629	218.94	
071123_0_R00X156Y092	13.5108	10.9126	18.7073	109.646	49.3665	72.231	94.0561	71.7113	144.462	183.955	
071123_0_R00X156Y094	22.2561	19.4741	28.7474	52.8582	107.571	31.5294	124.263	115.917	236.471	204.014	
071123_0_R00X156Y096	25.5867	33.4596	14.7816	81.6807	101.363	81.6807	115.14	66.9191	84.633	106.283	
071123_0_R00X158Y092	15.8937	16.4613	24.9757	44.2752	22.1376	47.1133	64.1423	97.6325	186.75	221.944	
071123_0_R00X158Y094	7.40787	35.9811	17.9905	67.7291	135.458	66.6708	58.2047	180.964	155.565	139.691	
071123_0_R00X158Y096	9.55467	22.0492	21.6817	102.896	49.9783	66.8827	92.9743	113.921	153.61	207.63	
071123_0_R00X158Y098	23.4534	48.7831	28.1441	77.8653	101.319	51.5975	60.9789	58.1645	121.958	174.493	
071123_0_R00X158Y100	21.5343	3.58905	19.1416	45.4614	82.5483	25.1234	65.7993	98.53	203.38	197.398	
071123_0_R00X158Y102	23.4076	12.8742	31.6003	98.312	80.7563	80.7563	80.7563	102.994	122.89	152.15	
071123_0_R00X158Y104	14.6501	19.8208	13.7884	102.551	91.3479	38.7798	130.128	103.413	175.802	162.875	
071123_0_R00X160Y092	8.0751	19.9186	24.7636	99.0546	39.2988	36.6071	97.4395	80.751	145.89	123.818	
071123_0_R00X160Y094	15.9116	26.2887	30.4396	51.8857	80.2498	87.1679	62.9546	65.7218	214.461	150.814	
071123_0_R00X160Y096	8.60331	24.7345	54.8461	88.1839	74.2035	61.2986	74.2035	88.1839	176.368	153.784	
071123_0_R00X160Y098	17.4303	24.4024	43.9243	85.757	82.271	59.9602	64.8407	80.8766	138.745	119.223	
071123_0_R00X160Y100	11.8858	27.5921	38.0535	108.671	52.6373	78.1069	78.5314	156.638	115.887	128.622	
071123_0_R00X160Y102	15.9645	29.6484	19.3855	37.6307	90.0855	86.6646	95.7871	115.173	117.453	176.75	
071123_0_R00X160Y104	7.84482	16.25	23.5345	37.5431	59.3965	27.4569	45.9482	52.6724	75.6465	85.7327	
071123_0_R00X160Y106	14.4514	18.8497	31.4162	83.567	42.0977	70.3722	71.0006	59.6907	106.815	140.116	
071123_0_R00X162Y092	11.268	45.877	41.8527	145.68	47.4867	90.9492	111.071	100.607	174.655	143.265	
071123_0_R00X162Y094	23.7069	25.6824	27.658	84.9496	119.851	65.1939	55.9745	36.2188	69.145	67.8279	
071123_0_R00X162Y096	14.8063	14.8063	23.6901	75.5123	44.419	53.3028	69.5898	105.125	156.947	198.405	
071123_0_R00X162Y098	8.33317	17.3074	29.4866	93.5879	44.8709	46.794	98.075	70.5114	149.356	190.381	

Sheet1 / Sheet2 / Sh

Fig. 10.7 Example representation of a peak list generated by Spec Align and Microsoft Excel software. Each column represents a spectrum and each line represents peak intensity [7]

Component Score					
	PC1	PC2	PC3	PC4	PC5 ···
Spectrum (X058Y104)	-10.13370794	-0.445522876	-2.694187973	3.505954923	-1.365223441 ···
Spectrum (X058Y105)	-9.229694994	-0.567799789	0.237015377	1.116560832	-1.233019483 ···
Spectrum (X058Y106)	-5.602620225	-0.656359247	1.143176137	-0.865466241	-0.732606489 ···
Spectrum (X058Y107)	-6.270811331	-2.03169119	-0.285734608	-1.819995303	-0.650038451 ···
Spectrum (X058Y108)	-7.66653915	0.364231772	-0.655977851	3.500619798	-0.957857607 ···
Spectrum (X058Y109)	-5.794829016	-0.370002515	1.224927625	2.69823407	-0.576784283 ···
Spectrum (X058Y110)	-6.095335164	-0.961580719	1.025728904	2.007179238	-0.679957134 ···
Spectrum (X058Y111)	-10.74294884	-0.105700353	-1.750048371	3.79086667	-1.19753972 ···
Spectrum (X058Y112)	-7.764753996	-0.612978725	1.471680849	0.700552849	-0.801859914 ···
Spectrum (X058Y113)	8.627123127	-4.905947292	-0.942587556	-1.087932962	-6.388071894 ···
Spectrum (X058Y114)	⋮	⋮	⋮	⋮	⋮

Fig. 10.8 Example representation of a PCA result performed by Excel add-ins. Component scores are calculated for each mass spectrum, according to each principal component

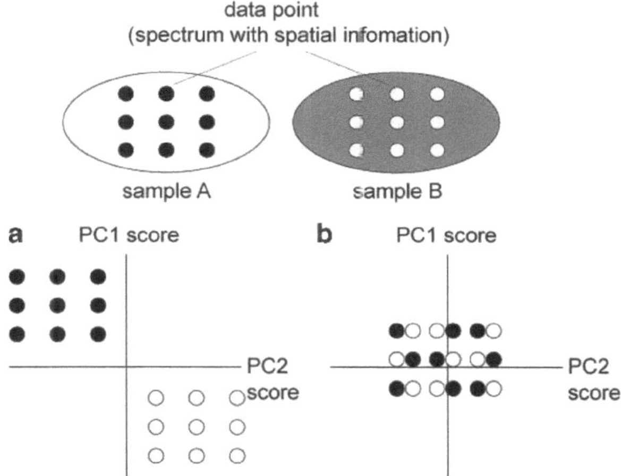

Fig. 10.9 Example of data interpretation of IMS-linked PCA. In this study, dots seen in the 2D plot represent the *case*, i.e., spectrum from distinct data points. If dots from distinct sample are separated (**a**), it means that the molecular expression patterns of these two regions were statistically distinct from each other. If not, PCA failed to extract the statistical differences between the two populations (**b**)

Note: We additionally noted that other artificial factors, such as variations in matrix application procedures, could be reflected as the main difference between the two groups in PC1. To avoid this, experiments should be performed in as identical conditions as possible (vis-à-vis matrix application, IMS measurement, etc.).

10.4.3.2   Interpretation of Factor Loading

An analysis of factor loading plot would identify peaks that were differentially expressed between two samples. A component score defined for each spectrum is a sum of the factor loading value, multiplied by each peak intensity. Therefore, when

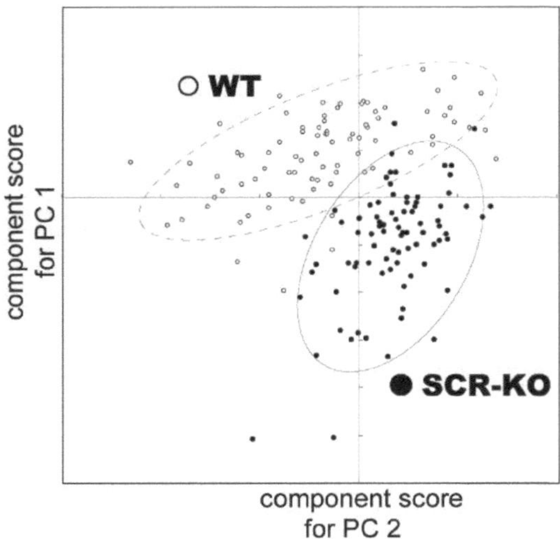

Fig. 10.10 Component scores for PC1 and PC2 are plotted on the *x*-axis and *y*-axis, respectively. Each spot represents a mass spectrum. *Open* and *filled spots* indicate spectra obtained from *WT* and *SCR-KO* mice, respectively

	A	PC1	PC2	PC3	PC4	PC5	...
1	**Factor Loading**						
2		PC1	PC2	PC3	PC4	PC5	...
3	m/z 1119	0.071506233	-0.005891377	-0.326683397	0.292888681	0.198868401	...
4	m/z 2153	-0.1486389	0.223434933	-0.489005528	0.230359915	0.091340758	...
5	m/z 9756	0.182920729	-0.152870098	-0.451659506	0.269897412	0.144066523	...
6	m/z 9809	-0.333347528	0.096143037	-0.345039238	0.089061045	0.062427567	...
7	m/z 10082	-0.294288151	-0.002173624	-0.418962896	-0.086867098	-0.038617682	...
8	⋮	⋮	⋮	⋮	⋮	⋮	
9	eigenvalue	17.89857465	4.949642519	4.509393439	2.764140748	2.075291342	
10	proportion	22.09700574	6.110669777	5.567152394	3.412519442	2.562088077	
11	cumulative proportion	22.09700574	28.20767551	33.77482791	37.18734735	39.74943543	

Fig. 10.11 Factor loading value is calculated for each mass peak

numbers ($=m$) of mass peaks were used in the analysis, the component score will be as follows:

$$\text{ScorePC1}(x, y) = \sum_{n=1}^{m} \text{load}(n) \times \text{Int.}(n),$$

where $\text{ScorePC1}(x, y)$ is the component score against PC1, obtained from (x, y); $\text{load}(n)$ is the factor loading value against a mass peak for n; $\text{Int.}(n)$ is the mass peak intensity for n; and m is the number of mass peaks used for calculation.

According to this equation, in the spectra from Scr-KO mice, the mass peak with large negative value regarding PC2 factor loading was supposed to be intense. On the other hand, in the WT sample, it was supposed that such peak intensities would be small. In other words, such mass peaks are suggested to differentiate samples derived from WT and Scr-KO tissue.

The factor loading value is calculated for each mass peak. In Fig. 10.11, the factor loading values for PC1 and PC2 are plotted on the *x*-axis and *y*-axis, respectively. Each spot indicates a distinct mass peak. Such a graph makes it very easy to find the peaks with the intended factor loading value against each PC. Because peaks that have negative loading values regarding PC2 are supposed to show different distributions between WT and Scr-KO, we picked up a mass peak at m/z 7,420 and obtained a distribution image (Fig. 10.12).

As a result, we found that this ion was highly expressed in the striatum of WT mice whereas it was expressed at remarkably low levels in the Scr-KO striatum (Fig. 10.13, arrowheads). Intriguingly, in olfactory bulb, there is no significant difference in the expression levels of m/z 7,420, between the two samples (Fig. 10.13, arrows). Furthermore,

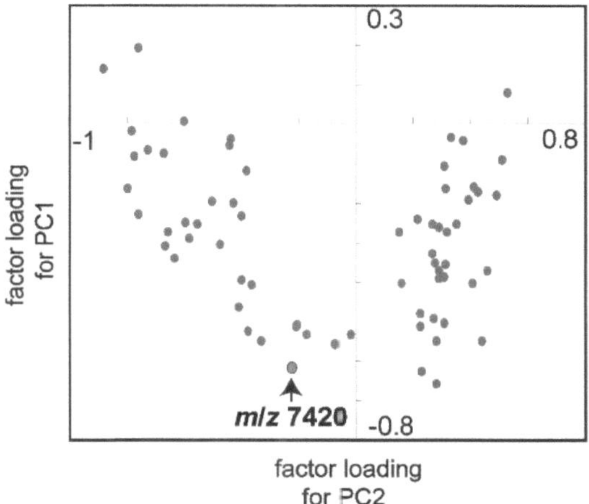

Fig. 10.12 Factor loading values for PC1 and PC2 are plotted on x-axis and y-axis, respectively. Each spot indicates the distinct mass peak. The peak (m/z 7,420) was chosen because it has large negative loading values regarding PC2

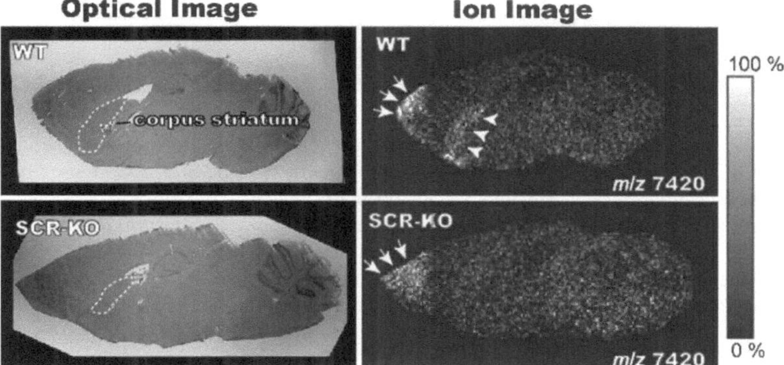

Fig. 10.13 H&E-stained brain sections prepared from WT and SCR-KO mice, with distribution of m/z 7,420. It was highly expressed in the striatum of WT mice but was expressed at remarkably low levels in Scr-KO striatum tissue (*arrowheads*). On the other hand, there was no significant difference in the expression levels of m/z 7,420 in the olfactory bulb, between the two samples (*arrows*) (Reprinted from Yao et al., Proteomics 8:3692–3701.)

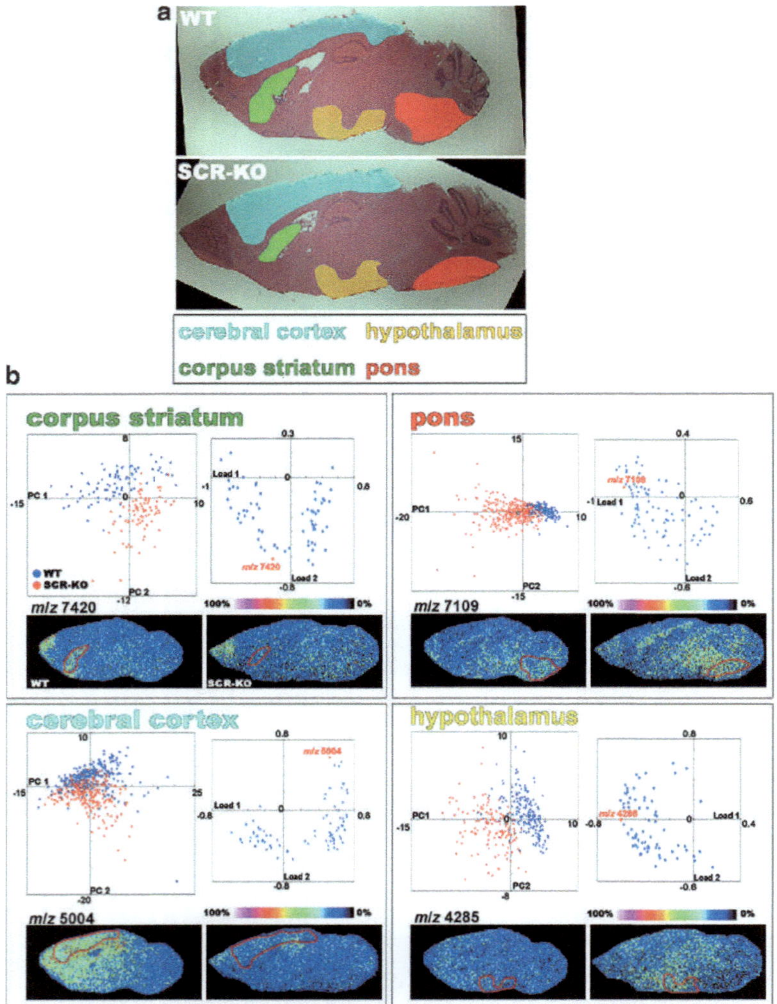

Fig. 10.14 PCA-linked IMS revealed abnormal expressions/suppressions of proteins in various regions of SCR-KO mouse brains. **a** H&E-stained images of WT and SCR-KO mouse brains. The regions focused upon during MS imaging analysis are indicated in *colors* as keyed below the pictures in (**a**). **b** Distributions of principal component scores of mass spectra, from various brain regions (*left spray graphs:* WT, *blue*; KO, *red*) and the factor loading plot (*right graphs*). The signal intensities of the mass spectra of the substances, with indicated *m/z*, are shown in the reconstructed images of the mouse brains analyzed with MS imaging [7] (Reprinted from Yao et al., Proteomics 8:3692–3701.)

through the same procedure, we were able to successfully identify the altered protein expression patterns in various selected brain regions (Fig. 10.14).

Note: We finally note that it is important to pay attention to the contribution of each PC to determine how many PCs are to be used in analysis. Proportion values represent the degree of a particular PC's contribution to the entire dataset (the values were seen in the 10th column of Fig. 10-10). For example, the proportion value of PC1 in this analysis is 22%, indicating that PC1 reflects 22% of the entire dataset. Any PC with too low a proportion value should be excluded from the analysis because we know which PC reflects data significantly in this manner. It is also important to note that PC1 often reflects variations in tissue sample preparation and matrix coating.

10.5 Conclusion

The volumes of IMS datasets continue to increase because of current improvements to IMS with regard to high resolution [20], three-dimensional (3D) imaging [21], and reconstruction from 3D mass spectra containing ion drift times in ion mobility MS [22]. Data analysis of such large datasets will increasingly depend on statistical analysis and will presumably be done automatically by software programs.

In this chapter, we showed a simple protocol that is needed to identify the differences between two samples. By using this minimal procedure, we were able to identify a molecule which differed between WT and Scr-KO tissue samples. Again, we would recommend that IMS experimenters perform statistical analysis by themselves, because undertaking such exercises will provide useful experience in understanding more complicated analyses, especially among biologists and clinicians.

References

1. Altelaar AF, Luxembourg SL, McDonnell LA, et al. (2007) Imaging mass spectrometry at cellular length scales. Nat Protocols 2:1185–1196
2. McCombie G, Staab D, Stoeckli M, et al. (2005) Spatial and spectral correlations in MALDI mass spectrometry images by clustering and multivariate analysis. Anal Chem 77:6118–6124
3. Plas RV, Moor BD, Waelkens E (2007) Imaging mass spectrometry-based exploration of biochemical tissue composition using peak intensity weighted PCA. In: 2007 IEEE/NIH Life Science Systems and Applications Workshop, pp 209–212
4. Yanagisawa K, Shyr, Y, Xu, B J, et al. (2003) Proteomic patterns of tumour subsets in non-small-cell lung cancer. Lancet 362:433–439
5. Mantini D, Petrucci F, Del Boccio P, et al. (2008) Independent component analysis for the extraction of reliable protein signal profiles from MALDI-TOF mass spectra. Bioinformatics 24:63–70
6. Prideaux B, Atkinson SJ, Carolan VA, et al. (2007) Sample preparation and data interpretation procedures for the examination of xenobiotic compounds in skin by indirect imaging MALDI-MS. Int J Mass Spectrom 260:243–251
7. Yao I, Sugiura Y, Matsumoto M, et al. (2008) In situ proteomics with imaging mass spectrometry and principal component analysis in the Scrapper-knockout mouse brain. Proteomics 8:3692–3701
8. Hanselmann M, Kirchner M, Renard BY, et al. (2008) Concise representation of mass spectrometry images by probabilistic latent semantic analysis. Anal Chem 80(24):9649–9658

9. Fowlkes EB, Mallows CL (1983) A method for comparing two hierarchical clusterings. J Am Stat Assoc 78:553–584
10. Deininger SO, Schürenberg M, Suckau D, et al. (2007) Class imaging: classification of breast cancer sections by MALDI tissue imaging. Poster presentation in HUPO
11. Denkert C, Budczies J, Kind T, et al. (2006) Mass spectrometry-based metabolic profiling reveals different metabolite patterns in invasive ovarian carcinomas and ovarian borderline tumors. Cancer Res 66:10795–10804
12. Lapolla A, Ragazzi E, Andretta B, et al. (2007) Multivariate analysis of matrix-assisted laser desorption/ionization mass spectrometric data related to glycoxidation products of human globins in nephropathic patients. J Am Soc Mass Spectrom 18:1018–1023
13. Wong WHJ, Cagney G, Cartwright HM (2005) SpecAlign: processing and alignment of mass spectra datasets. Bioinformatics 21:2088–2090
14. Yao I, Takagi H, Ageta H, et al. (2007) SCRAPPER-dependent ubiquitination of active zone protein RIM1 regulates synaptic vesicle release. Cell 130:943–957
15. He L, Lu XY, Jolly AF, et al. (2003) Spongiform degeneration in mahoganoid mutant mice. Science 299:710–712
16. Sugiura Y, Shimma S, Setou M (2006) Thin sectioning improves the peak intensity and signal-to-noise ratio in direct tissue mass spectrometry. J Mass Spectrom Soc Jpn 54:4
17. Schwartz SA, Reyzer ML, Caprioli RM (2003) Direct tissue analysis using matrix-assisted laser desorption/ionization mass spectrometry: practical aspects of sample preparation. J Mass Spectrom 38:699–708
18. Norris JL, Cornett DS, Mobley JA, et al. (2006) Processing MALDI mass spectra to improve mass spectral direct tissue analysis. Int J Mass Spectrom 260:212–221
19. Sugiura Y, Konishi Y, Zaima N, et al. (2009) Visualization of the cell-selective distribution of PUFA-containing phosphatidylcholines in mouse brain by imaging mass spectrometry. J Lipid Res (in press)
20. McDonnell LA, Piersma SR, Maarten Altelaar AF, et al. (2005) Subcellular imaging mass spectrometry of brain tissue. J Mass Spectrom 40:160–168
21. Andersson M, Groseclose MR, Deutch AY, et al. (2008) Imaging mass spectrometry of proteins and peptides: 3D volume reconstruction. Nat Methods 5:101–108
22. McLean JA, Ridenour WB, Caprioli RM (2007) Profiling and imaging of tissues by imaging ion mobility-mass spectrometry. J Mass Spectrom 42:1099–1105

Chapter 11
Statistical Analysis of IMS Dataset with ClinproTool Software

Nobuhiro Zaima and Mitsutoshi Setou

Abstract The procedures of performing principal component analysis (PCA) (1) and data mining method for biomarker discovery (2) using ClinProTools 2.1 software (Bruker Daltonics) are described. (1) First, PCA of an IMS dataset is described. PCA is widely used in many fields, such as biology, medicine, pharmacy, and economics. PCA is a statistical method used to reduce multidimensional data sets to lower dimensions, and the results of PCA are generally discussed in terms of component scores and loadings (see Sect. 11.1). (2) Second, the data mining method from an IMS dataset is described. Biomarkers are objective indicators of particular pathogenic processes, pharmacological responses, or normal biological states and can be any kind of molecule in living organisms. Biomarkers are essential for the diagnosis and prediction of diseases, for example, mass screening for newborn infants using biomarkers is performed in many countries. The accurate and effective processing of complicated data is needed for biomarker discovery (see Sect. 11.2).

11.1 PCA of IMS Dataset with ClinProTools Software

PCA is a statistical method used to reduce multidimensional datasets to lower dimensions. PCA is used as a data mining tool in many fields, including biology, medicine, pharmacy, and economics. PCA results are generally discussed in terms of component scores and loadings. The procedure in performing PCA, using ClinProTools 2.1 (Bruker Daltonics) is introduced in this section.

N. Zaima and M. Setou (✉)

Department of Molecular Anatomy, Hamamatsu University School of Medicine,
1-20-1 Handayama, Higashi-ku, Hamamatsu, Shizuoka 431-3192, Japan
e-mail: setou@hama-med.ac.jp

M. Setou (ed.), *Imaging Mass Spectrometry: Protocols for Mass Microscopy*,
DOI 10.1007/978-4-431-09425-8_11, © Springer 2010

11.1.1 PCA Protocol

1. *Peak processing*

In this step, performing baseline correlations, choosing peaks, and recalibrating
for PCA are performed:

(a) Load data
   Start ClinProTools 2.1 and select a data folder by clicking "File→Open
   Model Generation Class" (Fig. 11.1).

2. *PCA*

Open the PCA result window (Fig. 11.2). Scores are shown in the upper part and
the Loading Plot is shown in the lower part of the window.

The meanings of the displayed figures are as follows (from left, sequentially): first,
the three-dimensional (3D) plot of the first, second, and third principal components
(upper, score plot; lower, loading plot); second, the two-dimensional (2D) plot of the
first and second principal components (upper, score plot; lower, loading plot); third, the
2D plot of the first and third principal components (upper, score plot; lower, loading plot);

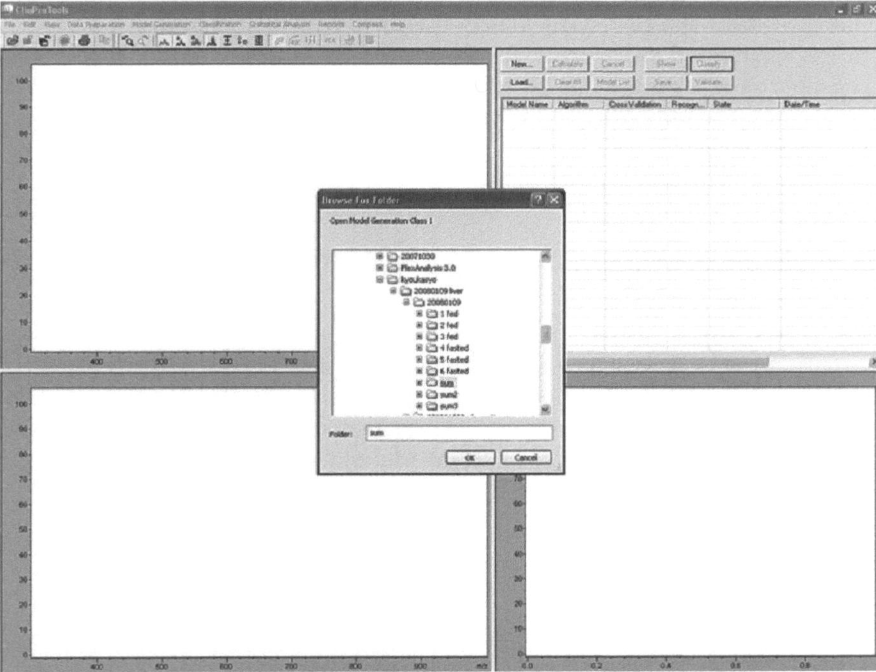

Fig. 11.1 ClinProTools 2.1 window. It is not necessary that the loading data be saved in one folder;
ClinProTools 2.1 can load data that have been saved to multiple folders. Data in other folders are
color-coded, following principal component analysis (PCA)

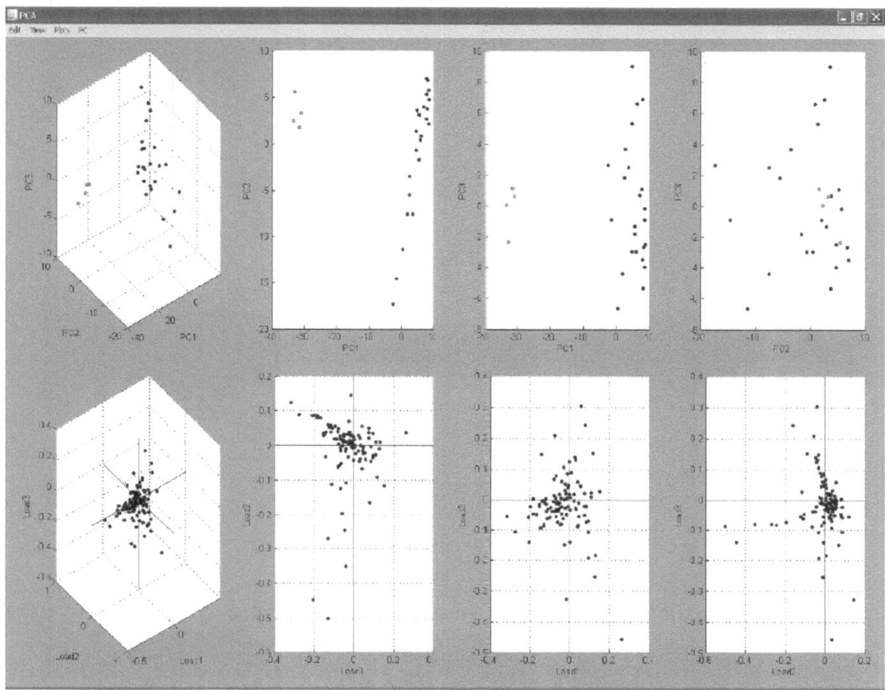

Fig. 11.2 ClinProTools 2.1 window displayed after PCA. Each plot can be displayed individually

fourth, the 2D plot of the second and third principal components (upper, score plot; lower, loading plot).

The peaks important for the separation of data are shown on the loading plot (see Fig. 11.2); the peaks that have large absolute values strongly influence the separation of data. When an enlargement of a specific plot needs to be shown, select the specific plot from "Plot" in the menu bar; the first, second, and third principal components are presented in the default window. Other principal components (e.g., fourth or fifth principal components) can also be displayed through this software.

The direct mass spectrometry (MS) data resulting from the PCA analysis of disordered tissues are shown in Fig. 11.3. The 2D plot (first and second principal components) is displayed. It was found that datasets were classified into three groups (see Fig. 11.3; circles added by the author). A loading plot is shown in Fig. 11.4, wherein the m/z values are displayed by clicking each data point. Several peaks (e.g., m/z 844.44) are characteristically different among datasets. Both the proportion rate and cumulative proportion rate can be verified by clicking "Plot→Variance" (Fig. 11.5).

3. *Representation of calculation values*

The PCA calculation values can be displayed by selecting "Files→ClinProtPCA" in the ClinProTools folder. The required calculation values need to be saved, because calculation values in ClinProtPCA are overwritten whenever PCA is performed.

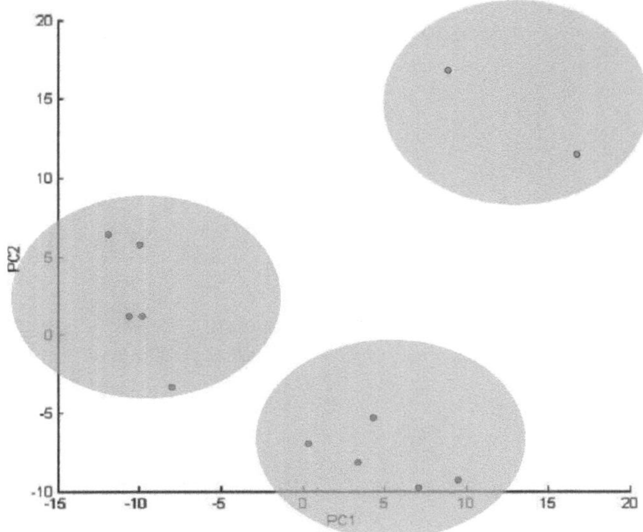

Fig. 11.3 Two-dimensional plot of the first and second principal components. The analyzed data are classified into three groups. It is suggested that the peak patterns of adjacently plotted datasets resemble each other

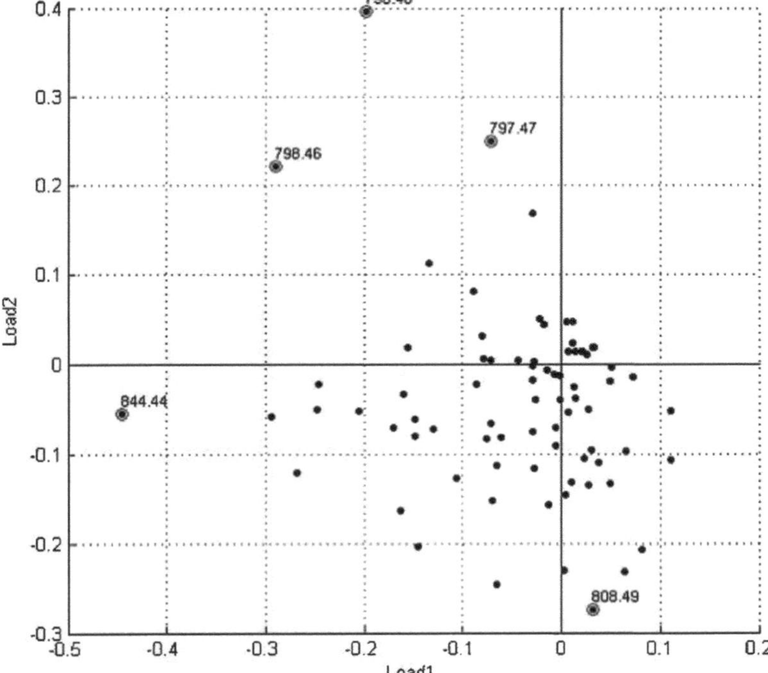

Fig. 11.4 Two-dimensional load plot

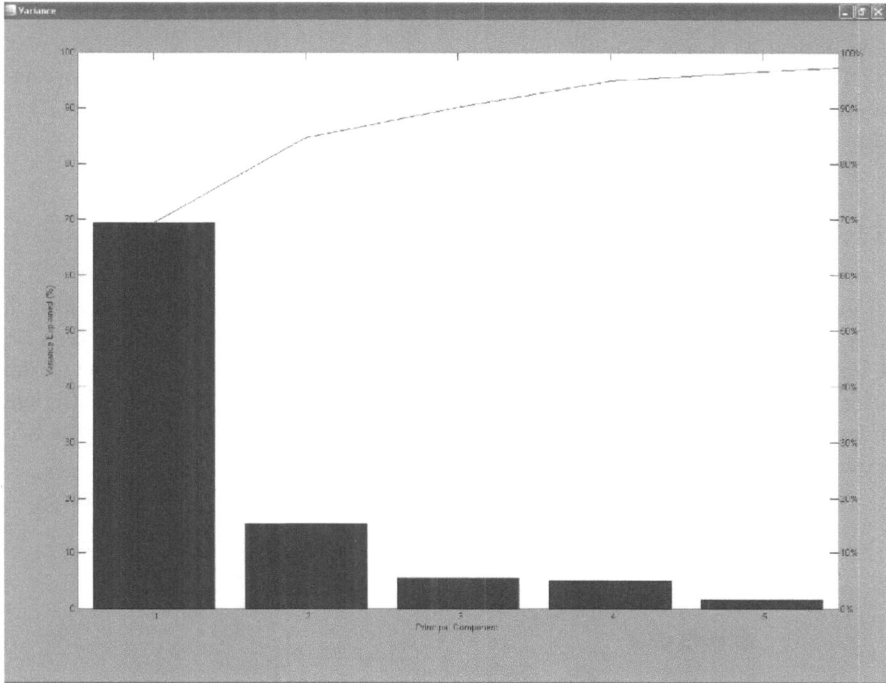

Fig. 11.5 Proportion rate and cumulative proportion rate

Note: As a software package, ClinProTools 2.1 is designed to analyze the data of the protein region that are acquired in linear mode. The PCA data demonstrated in the figures in this section pertain to a low-weight molecule acquired in the reflector mode of ClinProTools 2.1; as such, the isotope peak is not considered in the software algorithm. Whenever data involve the analysis of a protein region in a mode other than linear mode, attention needs to be paid to accurate data interpretation.

11.2 Biomarker Discovery from IMS Dataset with ClinProTools

Biomarkers are objective indicators of particular pathogenic processes, pharmacological responses, or normal biological states; they can involve any kind of molecule in living organs, e.g., lipids, proteins, peptides, or DNA. Biomarkers are essential for the diagnosis and prediction of diseases; for example, in many countries, mass screening of newborn infants is performed using biomarkers. The use of biomarkers in drug development or the creation of made-to-order medicine is also anticipated. IMS can provide distribution information regarding various biomolecules at the cell and tissue levels, and so it is expected to become a powerful tool for in situ biomarker discovery.

The accurate and effective processing of complicated data is needed for biomarker discovery. In this section, a data mining method using ClinProTools 2.1 is described.

11.2.1 Biomarker Discovery by Comparing Two Groups

A biomarker discovery method executed by comparing two groups is described in this section (Fig. 11.6). First, find biomarker candidates by using training data (in ClinProTools software, biomarkers are called "models"). Second, assess the model using test data, to investigate whether the established model provides sufficient reliability. A model consists of several peaks that are needed to distinguish the two groups.

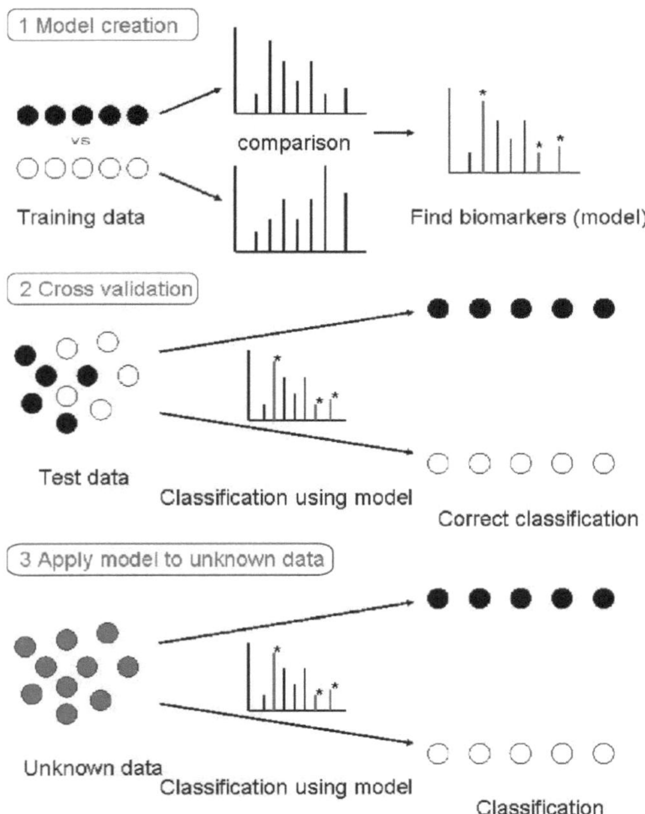

Fig. 11.6 Biomarker discovery

11.2.2   *Procedure for Biomarker Discovery*

The following list details, step by step, the procedure that one must take to discover biomarkers using ClinProTools 2.1:

1. *Sample data acquisition*

MS data of each sample are acquired using direct MS techniques. MS data obtained from the same group need to be saved in the same folder, for the next step.

2. *Peak processing*

In this step, the correction of baseline correlations, the normalization of peak intensities, choosing peaks, and performing recalibration are undertaken before comparing the MS data of each sample:

(a) Data loading
   Start ClinProTools 2.1 (Fig. 11.7) and load MS data by clicking "File→Open Model Generation" or 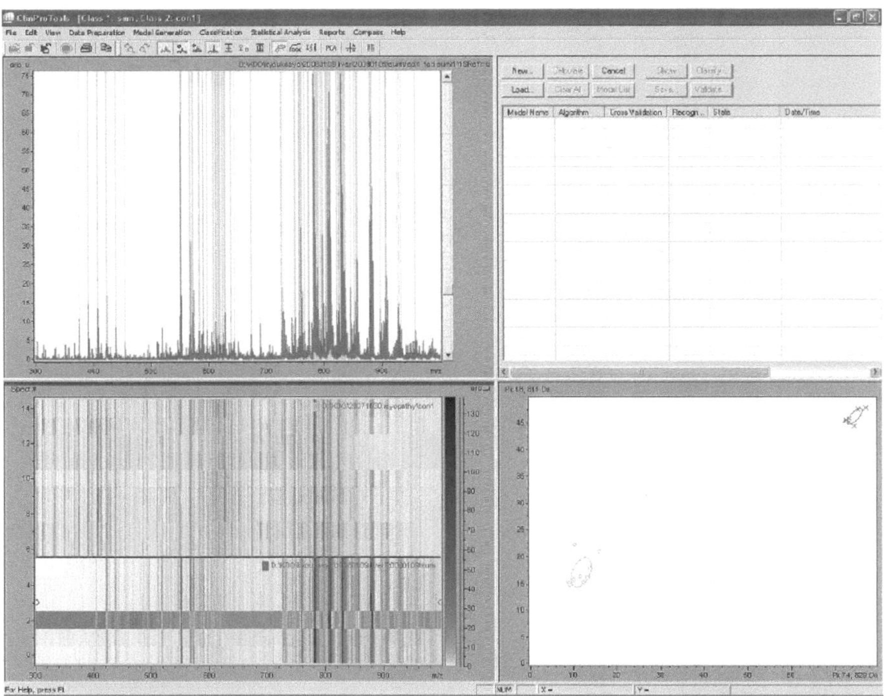 .

Fig. 11.7 ClinProTools 2.1 window. Each spectrum can be enlarged in the "Spectrum View" window

(b) Peak processing

Start peak processing by clicking "Reports→Peak Statistic" or [icon]. After peak processing, the chosen peak areas become blue colored in the "Spectrum View" window (see Fig. 11.7, left upper) and 2D data are presented (Fig. 11.7, right lower). The two axes of these data are two characteristic peaks that are suitable for separating two groups, and the peaks of the two axes can be selected manually. Unprocessed data are red colored (Fig. 11.7, left lower), and thus it is essential to verify whether all peak processing is performing normally. The ClinProTools "Peak Statistic" window, which shows the chosen peaks together with statistical data, is also shown (Fig. 11.8).

3. *Model generation*

A model is created after peak processing. An assessment of model reliability (i.e., cross-validation) can be performed simultaneously. In this software, to perform cross-validation, each group must contain a minimum of 20 records.

(a) Select algorithm for model creation

Select algorithm for model creation by clicking "Model Generation→New Model" or [New...], and name the new generation model. Four kinds of

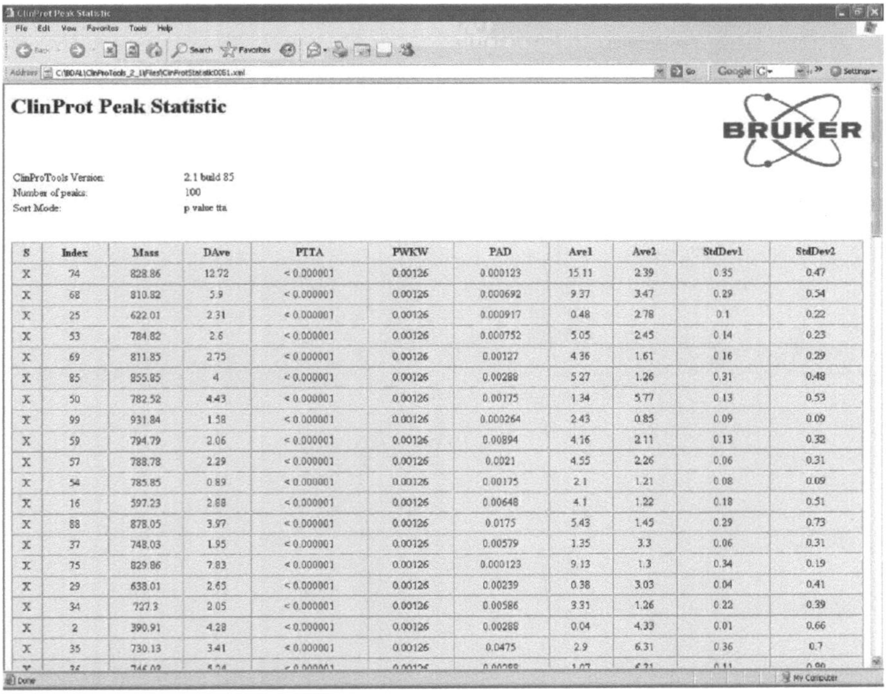

ClinProt Peak Statistic

BRUKER

ClinProTools Version:	2.1 build 85
Number of peaks:	100
Sort Mode:	p value tta

S	Index	Mass	DAve	PTTA	PWKW	PAD	Ave1	Ave2	StdDev1	StdDev2
X	74	828.86	12.72	< 0.000001	0.00126	0.000123	15.11	2.39	0.35	0.47
X	68	810.82	5.9	< 0.000001	0.00126	0.000692	9.37	3.47	0.29	0.54
X	25	622.01	2.31	< 0.000001	0.00126	0.000917	0.48	2.78	0.1	0.22
X	53	784.82	2.6	< 0.000001	0.00126	0.000752	5.05	2.45	0.14	0.23
X	69	811.85	2.75	< 0.000001	0.00126	0.00127	4.36	1.61	0.16	0.29
X	85	855.85	4	< 0.000001	0.00126	0.00288	5.27	1.26	0.31	0.48
X	50	782.52	4.43	< 0.000001	0.00126	0.00175	1.34	5.77	0.13	0.53
X	99	931.84	1.58	< 0.000001	0.00126	0.000264	2.43	0.85	0.09	0.09
X	59	794.79	2.06	< 0.000001	0.00126	0.00894	4.16	2.11	0.13	0.32
X	57	788.78	2.29	< 0.000001	0.00126	0.0021	4.55	2.26	0.06	0.31
X	54	785.85	0.89	< 0.000001	0.00126	0.00175	2.1	1.21	0.08	0.09
X	16	597.23	2.88	< 0.000001	0.00126	0.00648	4.1	1.22	0.18	0.51
X	88	878.05	3.97	< 0.000001	0.00126	0.0175	5.43	1.45	0.29	0.73
X	37	748.03	1.95	< 0.000001	0.00126	0.00579	1.35	3.3	0.06	0.31
X	75	829.86	7.83	< 0.000001	0.00126	0.000123	9.13	1.3	0.34	0.19
X	29	638.01	2.65	< 0.000001	0.00126	0.00239	0.38	3.03	0.04	0.41
X	34	727.3	2.05	< 0.000001	0.00126	0.00586	3.31	1.26	0.22	0.39
X	2	390.91	4.28	< 0.000001	0.00126	0.00288	0.04	4.33	0.01	0.66
X	35	730.13	3.41	< 0.000001	0.00126	0.0475	2.9	6.31	0.36	0.7
X	36	746.03	5.24	< 0.000001	0.00126	0.00288	1.07	6.31	0.11	0.90

Fig. 11.8 The ClinProTools "Peak Statistic" window. Statistical data of processed peak are shown in this window

algorithm [i.e., Support Vector Machine (SVM), Genetic Algorithm (GA), Supervised Neural Network (SNN), and Quick Classifier (QC)] are available for model generation in ClinProTools 2.1.

(b) Start calculations for model generation

Start calculations by clicking "Model Generation→Calculation" or [Calculate]. The generated model then appears in the model list (Fig. 11.9, right upper).

(c) Show result

Right-click on the generated model in the model list and select "Show Model" to open the ClinProModel window; selected peaks (biomarkers) will be listed therein (Fig. 11.10). The color of the selected peak areas will change from light blue to pink (see Fig. 11.9, left upper).

(d) Save model

The generated model will not be saved automatically. To do so, right-click on the generated model and select "Save Model As" or click [Save...].

4. *Assessment of model reliability*

The reliability of the generated model is assessed, using test data. The data used for model generation cannot be used for assessment; two kinds of test data are required (e.g., normal group and disorder group):

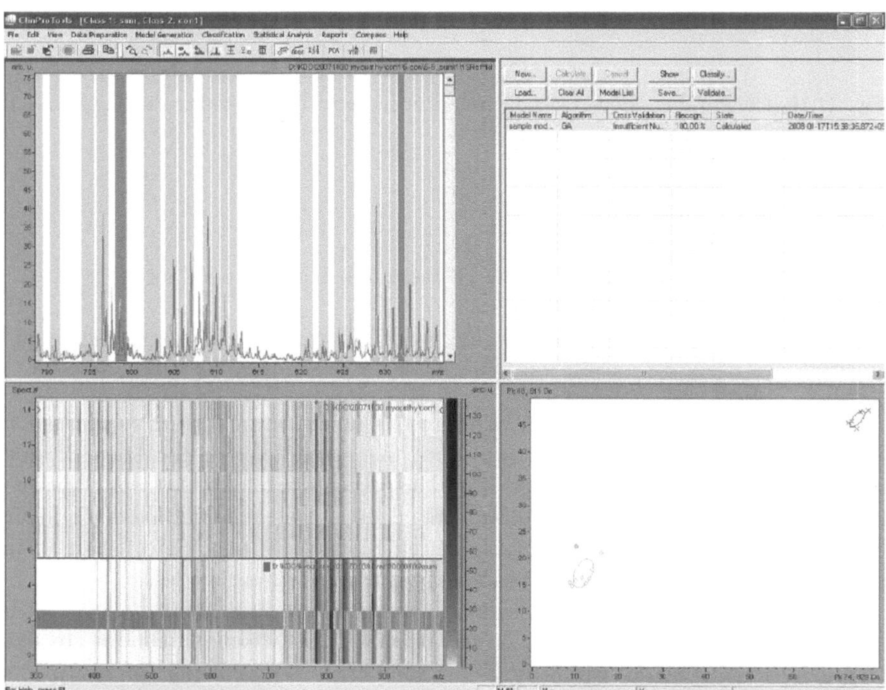

Fig. 11.9 The ClinProTools 2.1 window after model calculation. For the first analysis, it is recommended that all four algorithms be used for model generation

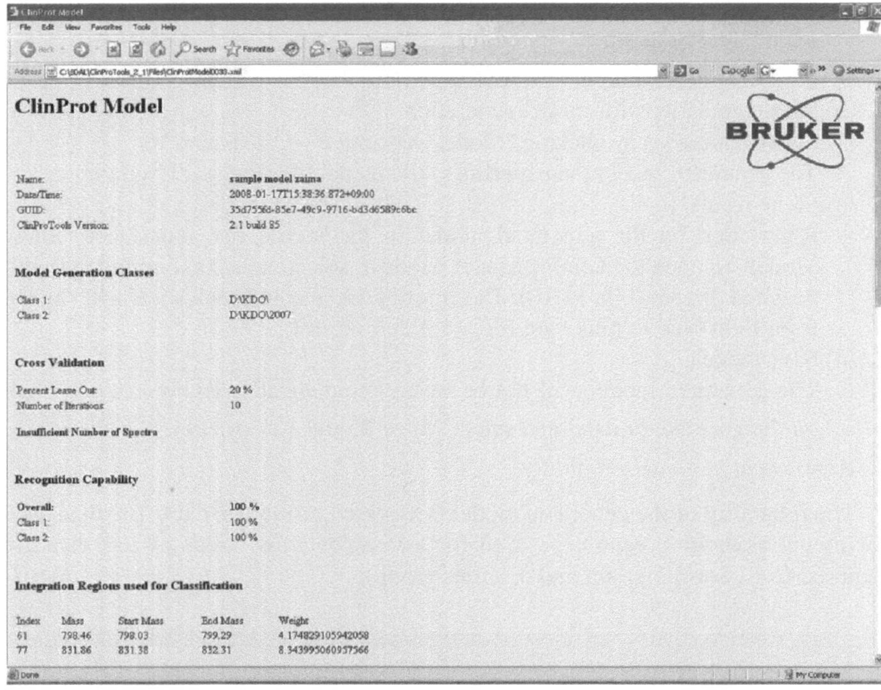

Fig. 11.10 ClinProTools Model window. Information about the biomarker candidate peak is displayed in this window

(a) Load saved model

    Load model by clicking "Model Generation→Load Model" or [Load...].

(b) Load test data

    Load two kinds of test data by clicking "Classification→External Validation" or [Validate...].

After the data are loaded, the ClinProtClassification and ClinProtValidation windows will be automatically displayed; the former shows the classification results of each dataset and the latter shows the percentage of data that was classified correctly (Fig. 11.11). Only when the generated model is deemed sufficiently reliable can the model be applied to the classification of unknown data. Selected peaks in the generated model (biomarkers) can be assigned via multistage MS.

Note: When there are many misclassifications of test data, another model must be generated with new algorithms or new training data. Misclassifications are also caused by heterogeneities within the biopsy sample or matrix coating. Improvements to the quality of MS data are important, if biomarker discovery is to proceed. The percentage of misclassifications can be decreased by changing the peak processing parameters of the software.

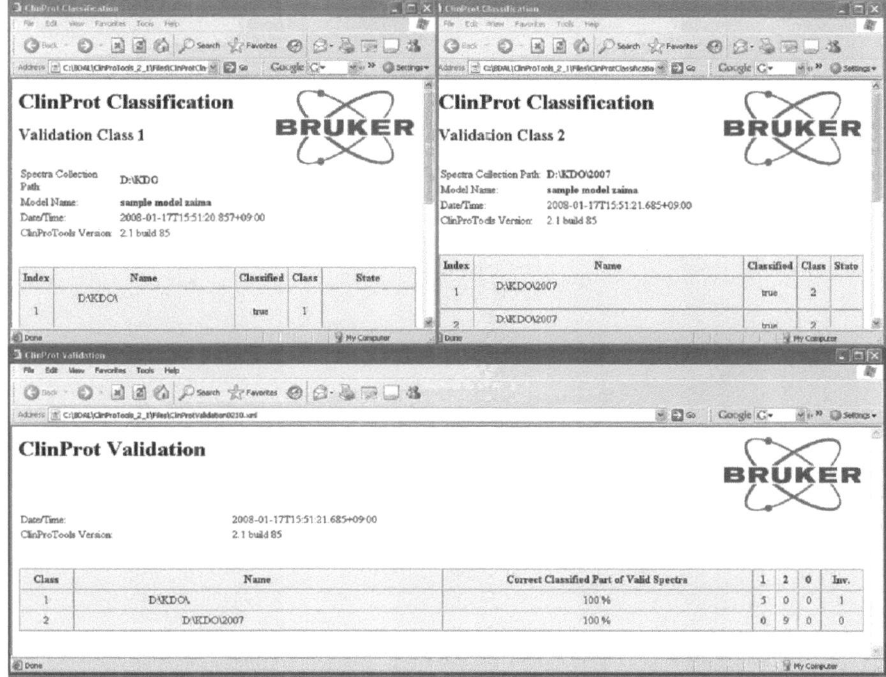

Fig. 11.11   ClinProTools Classification (*upper*) and ClinProTools Validation (*lower*). The classification percentages are shown in the ClinProTools Classification window

5. *Apply established model for new data*

When a reliable model is successfully established, it can be applied to new data. Load new data by clicking "Classification→Classify" or Classify... .

11.2.3   Correlations Between Peaks

Correlations between peaks can be displayed in two ways: through the Correlation Matrix view or the Correlation List view.

11.2.4   Correlation Matrix View

Display the Correlation Matrix view by clicking "Reports→CorrelationMatrix" or . A positive correlation is red colored and a negative correlation is blue colored (Fig. 11.12).

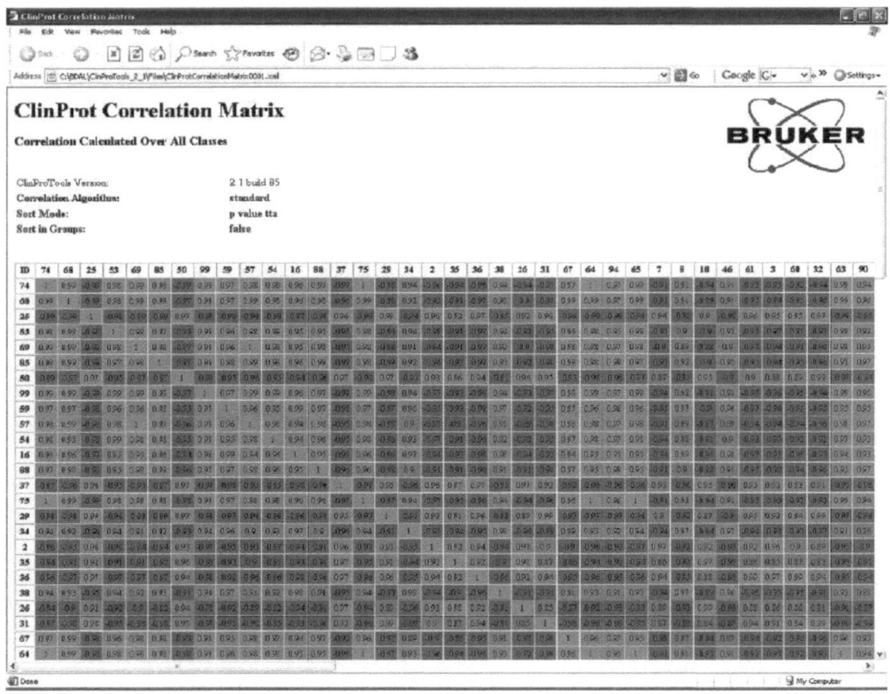

Fig. 11.12 The ClinProTools Correlation Matrix view. A positive correlation is red colored and negative correlation is blue colored

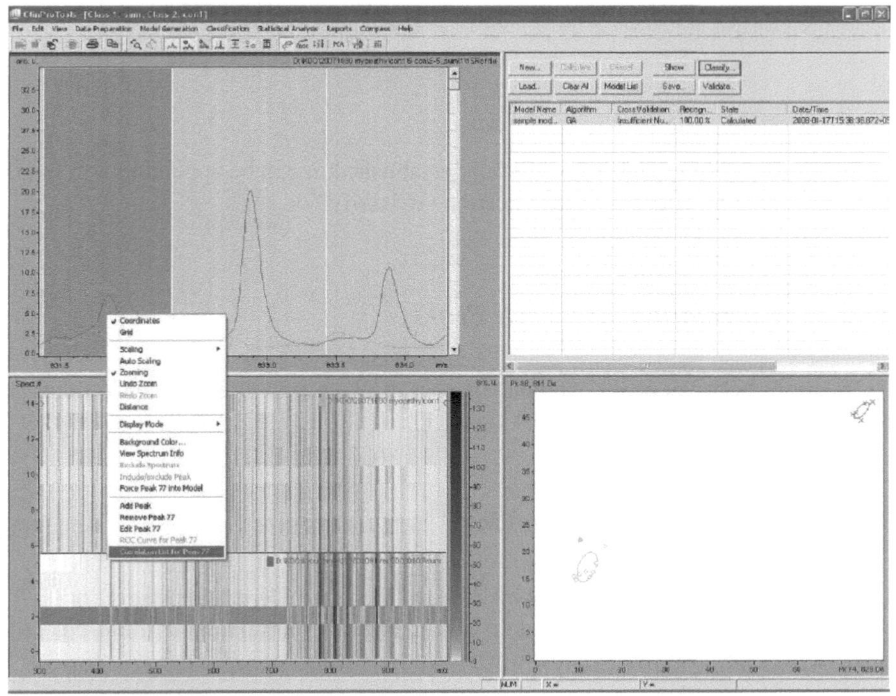

Fig. 11.13 Correlation list 1

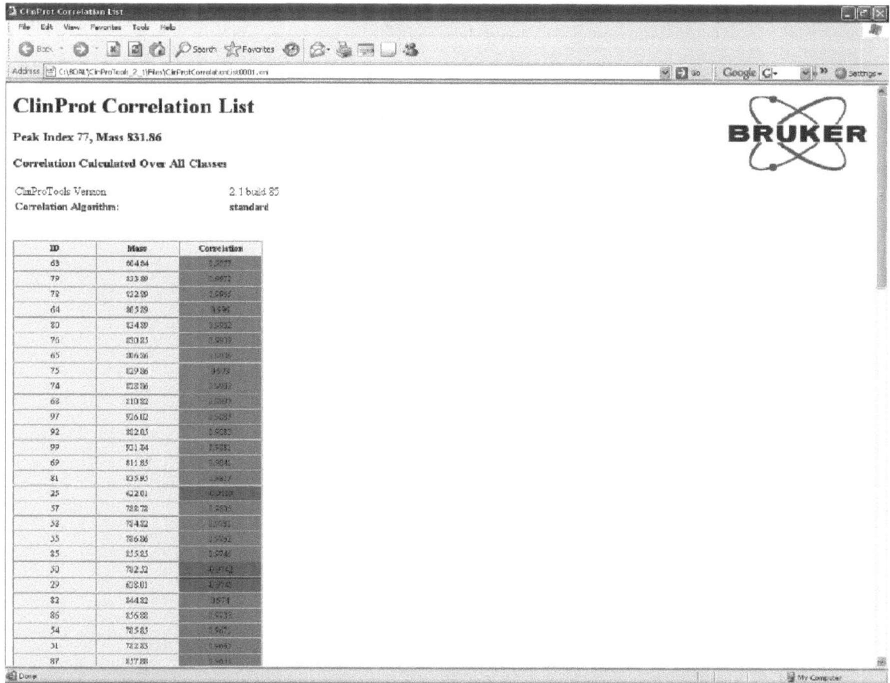

Fig. 11.14 Correlation list 2

11.2.5   Correlation List View

Right-click on the peak of interest in the spectrum view and select "Correlation List for Peak X" (Fig. 11.13), where "X" is the peak number (Fig. 11.14).

Note: The protocol described in this section is a basic procedure. There are many another functions available for biomarker discovery in ClinProTools.

Acknowledgments The author thanks Y. Matsuyama and T. Kudo, both of Bruker Daltonics, for their advice.

Part VI
Applications of MALDI-IMS

Chapter 12
Imaging of Cultured Cells by Mass Spectrometry

Hyun Jeong Yang, Yuki Sugiura, Koji Ikegami, and Mitsutoshi Setou

Abstract Establishment of an Imaging mass spectrometry (IMS) experimental procedure for cultured cells is an important issue because it provides information on localization of various types of biomolecules inside the cell. At present, how to prepare the samples from cultured cells for an IMS has been under study. In this section, we present the preparation method for IMS analysis of mouse superior cervical ganglion (SCG) explant culture, containing many sympathetic neurons. The SCG explant culture is much larger than a single cell and neurons are highly polarized cells, allowing the comparison of the distinct compartment of the cells by IMS. The quick freeze-dry method was applied for the fixation of the neurons, and distinct distributions of small metabolites in the neurons were successfully visualized. We found that molecular composition was remarkably different between the cell body-containing region and the axon-enriched region.

12.1 Introduction

As previous chapters of this book have described, imaging mass spectrometry (IMS) provides us with an attractive opportunity to observe distributions of various types of molecules in tissues. By applying it to cultured cells, we expect to obtain information of the molecular distribution within cellular organelles. At present, however, the laser spot diameter of the typical matrix-assisted laser desorption/ionization (MALDI) mass spectrometry (MS) instrument (10–100 µm) is still not enough for single cell imaging

H.J. Yang and Y. Sugiura
Department of Bioscience and Biotechnology, Tokyo Institute of Technology,
4259 Nagatsuta-cho, Midori-ku, Yokohama, Kanagawa 226-8501, Japan

H.J. Yang, Y. Sugiura, K. Ikegami, and M. Setou (✉)
Department of Molecular Anatomy, Hamamatsu University School of Medicine,
1-20-1 Handayama, Higashi-ku, Hamamatsu, Shizuoka 431-3192, Japan
e-mail: setou@hama-med.ac.jp

M. Setou (ed.), *Imaging Mass Spectrometry: Protocols for Mass Microscopy*,
DOI 10.1007/978-4-431-09425-8_12, © Springer 2010

[2, 3]. Generally, by use of the typical MALDI-time-of-flight (TOF) instrument, we could only obtain a few pixels from the single cell, which is not sufficient to visualize subcellular structures (Fig. 12.1).

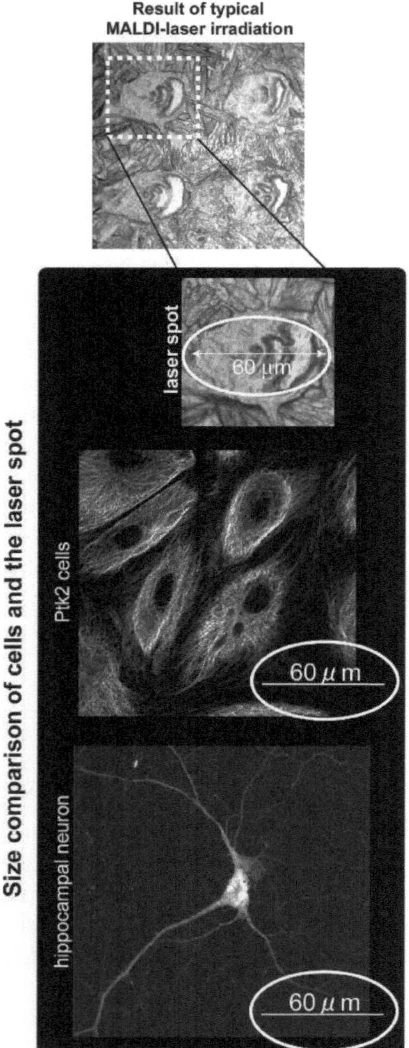

Fig. 12.1 Size comparison of typical laser spot diameter of matrix-assisted laser desorption/ionization (MALDI)-time-of-flight (TOF) mass spectrometry (MS) instrument and cultured cells. *Top panel:* Microscopic observation of "burned-out" region of tissue section coated with 2,5-dihydroxybenzoic acid (DHB) matrix. In this case, the diameter of the spot is approximately 60 μm. *Bottom panel:* Size comparison of the burned-out spot size and cultured cells, which are a Ptk2 cell and hippocampal neuron (Reprinted from Yang et al., Proceedings for 7th International Symposium on Atomic Level Characterizations for New Materials and Devices.)

In this regard, secondary ion mass spectrometry (SIMS)-based imaging can obtain much higher spatial resolution by electromagnetically focusing a high-energy ion beam (for details of the SIMS principle and its application, see Chap. 18). The imaging of cells with SIMS shows the subcellular distribution of atoms and small molecules including lipid metabolites [1, 4–8]. As shown in Fig. 12.2, Ostrowski et al. have nicely visualized that the molecular compositions in plasma membranes of a single *Tetrahymena* change remarkably, especially in adhering domain, during mating [1]. However, typical SIMS can ionize the $m/z < 1,000$ range of ions. In addition, it can hardly perform molecular identification by tandem mass spectrometry.

Thinking upside down, MALDI-IMS of relatively large-sized neurons is possible [2]. In this section, we used sympathetic neurons of the mouse superior cervical ganglion (SCG) for IMS, whose resolution is adequately fulfilled by the large-sized explant culture.

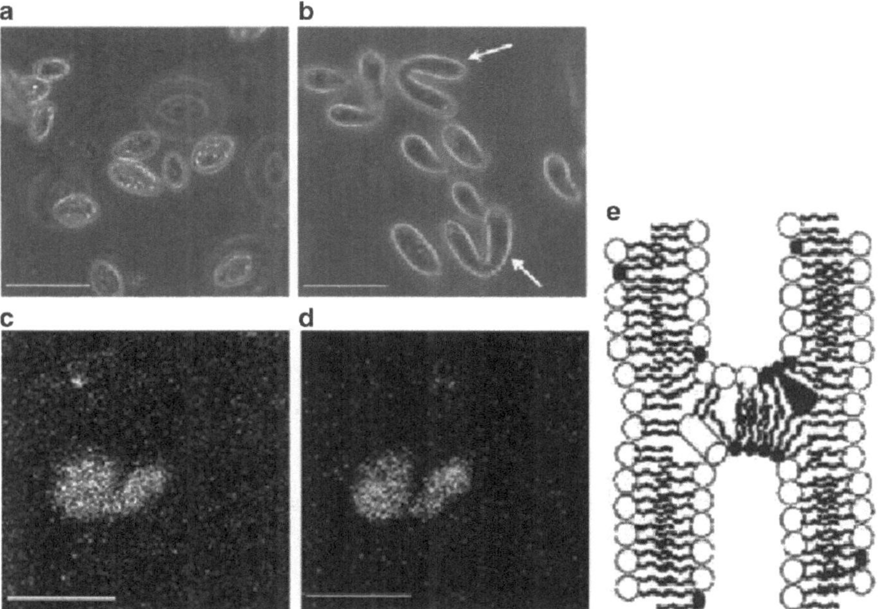

Fig. 12.2 Results of secondary ion mass spectrometry (SIMS) imaging of *Tetrahymena* during their conjugation. Optical images of *Tetrahymena* before mating (**a**) and during full conjugation (**b**) (*arrows*). **c** SIMS image for m/z 69 (C5H9$^+$). **d** SIMS image for m/z 184 (PC head group), demonstrating lipid heterogeneity at the conjugation junction. SIMS images were acquired with a lateral resolution of approximately 250 nm. *Bars* 50 μm. **e** A schematic of the membrane fusion intermediate structure, the stalk. The *wavy lines* depict the acyl tail groups of the membrane phospholipids. The *white circles* represent the head group of PC, a cylinder-shaped lamellar lipid. The *black circles* represent the head group of PE, a cone-shaped nonlamellar lipid. Membrane fusion sites probably contain a large quantity of cone-shaped lipids, because those lipids fit well into contoured intermediate structures. (Adapted from [1]) (Reprinted from Ostrowski et al., Science 305:71–73.)

12.2 Preparation of Cultured Cells for IMS

At present, how to prepare the samples from cultured cells for IMS still remains controversial. Although cell preparation protocols for traditional immunostaining and other dye staining are established, the method to prepare cultured cells for IMS is still on the way of development. There are several problems to be overcome. For example, cultured cell monolayers are structurally weaker than tissue slices; thus, how to fix cells for IMS is one of the essential problems. Reagents generally used for cell fixation such as paraformaldehyde (PFA) might cause protein/peptide signal reduction by the formation of cross-linkage among proteins. For small molecular analysis, loss or migration of small molecules from original location might occur to solve this issue. Development of alternative methods that achieve morphological preservation of cells are required. In Table 12.1, Rubakhin et al. studied several methods for cell fixation, which are air-drying, glycerol substitution, and oil substitution; they reported that glycerol substitution showed the best results [2].

In the example shown here, mouse neuronal cells were fixed by lyophilization, as is done in several reports regarding the IMS of cultured cells [4–8]. Matrix application should also be carried out with extreme care (preferably by the automatic or manual spray coating method), because a slight diffusion of analyte in small-sized samples can result in the detection of artificial (i.e., false) localizations in imaging. Therefore, matrix-free imaging of cells is an important issue to investigate, as a recent study reported a desorption/ionization on silicon (DIOS) method for single-cell IMS [9].

In this section, we present the IMS of SCG neurons. SCG, consisting of large (>20 μm in diameter) peripheral sympathetic neurons, is suitable for discriminating between cell bodies and neurites. We analyzed the distribution of the molecular composition

Table 12.1 Comparison of fixation methods in detection of proteins and peptides

Methods	Preservation of morphology	Quality of signals	Accessibility to analytes	Diffusion of analytes
Replaced by air	Bad	Bad	Bad	Not detected
Stabilized by glycerol	Good	Very good	Very good	Not detected
Substituted with FC-43, mineral oil	Bad	Moderate	Moderate	Occurs
Fixed by PFA	Excellent	Moderate	Moderate	Occurs

Reprinted with permission from Rubakhin et al. (2003) [2]; copyright 2006, American Chemical Society

of cultured SCGs. According to our IMS protocol, we found that molecular compositions were remarkably different between cell bodies and neurites.

12.2.1 Coating of ITO-Glass Slide

Metal-coated glass slides are popularly used in IMS [10–13]. It is convenient to culture cells on their metal-coated surface for IMS, as some cells can attach and grow on glass slides coated with indium thin oxide (ITO) [14–17]. In addition, poly-L-lysine (PLL) and laminin coating is an essential step for neural cell culture to elongate their neuronal processes; the following steps detail how to coat PLL and laminin on the ITO slide glass:

1. The ITO-coated glass slides are purchased from (Bruker Daltonics, or Sigma Aldrich) (Fig. 12-3a). The following procedures should be performed on a clean bench.
2. Irradiate UV on ITO-coated glass slides overnight, for sterilization.
3. Wash the slides with 70% ethanol, 100% ethanol for several hours, and then sterilized water. Allow them to air-dry.
4. Attach flexiPERM (Greiner Bio-One) to the slides (Fig. 12.3b,c).
5. Coat the surface of the slides with 1 mg ml^{-1} PLL in borate buffer at 37°C overnight.
6. Wash the slides with sterilized water once.
7. Coat the surface of the slides with 10 μg ml^{-1} laminin in minimum essential medium (MEM) at 37°C for 3 h.
8. Wash the slides with sterilized water three times.

12.2.2 Preparation of SCG Cell Culture

1. Anesthetize and decontaminate 0- or 1-day-old mouse pups by covering them with 70% ethanol solution. Do not continue this step for more than 10 min, because longer anesthetization causes damage to neuronal cells, resulting in a lower survival rate of extracted ganglia.
2. Decapitate the pups from the dorsal side with scissors. Make sure to cut near the shoulder, to avoid leaving ganglia in the body.
3. Hold the head face-up on a glass Petri dish filled with congealed paraffin. The dish is cooled by being placed on ice.
4. Clean away blood by adsorbing with cotton-tipped swabs.
5. Bring out the SCG from the cervical region. The ganglion is found behind the branching point of the carotid artery, after removing the neck muscles.
6. Remove the carotid artery and connective tissues from the SCG. This step to clean up the ganglion is important for making a good explant culture.
7. Divide the ganglion into halves, using 18-gauge needles.

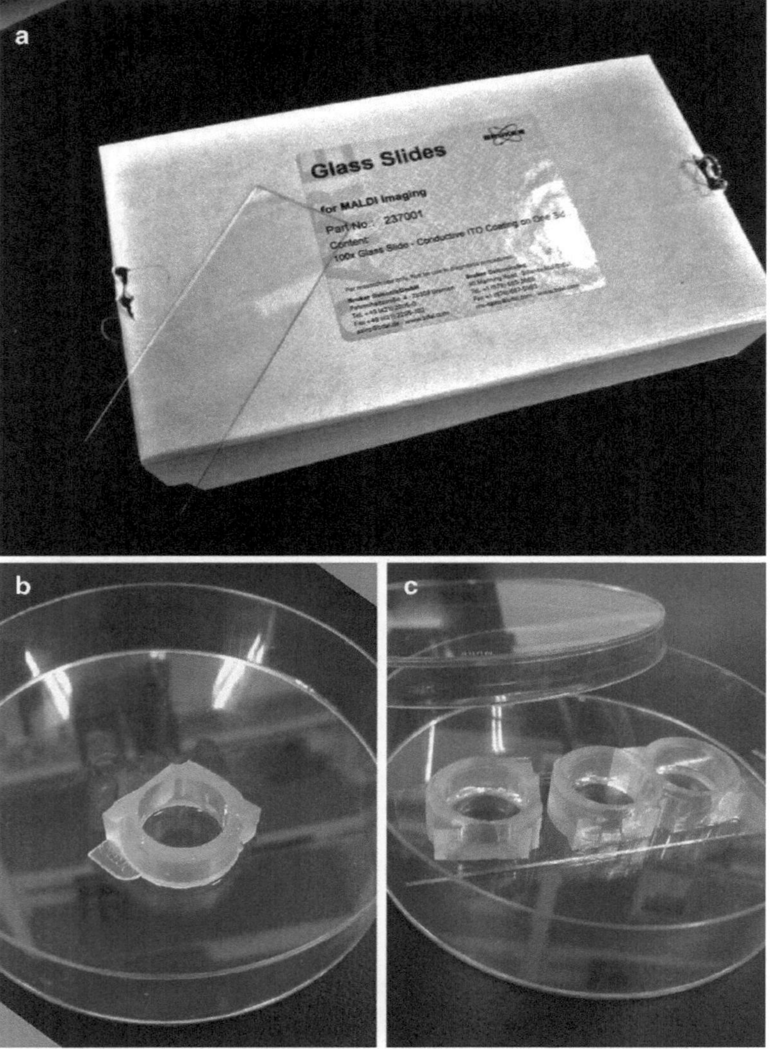

Fig. 12.3 ITO-coated glass slides (**a**) and flexiPERM (**b**). Attach flexiPERM to the glass slides and place them on a 10-cm Petri dish under a clean environment (**c**) (Adapted from [18]) (Reprinted from Yang et al., Proceedings for 7th International Symposium on Atomic Level Characterizations for New Materials and Devices.)

8. Place a piece of half a ganglion at the center of a flexiPERM chamber that is filled with 160 µl AM50 medium containing MEM supplemented with 50 ng ml^{-1} nerve growth factor, 15 µM fluorodeoxyuridine, and 10% fetal bovine serum. Incubate the cell culture at 37°C in a humidified incubator with 5% CO_2/95% air.
9. To eliminate nonneuronal cells, add an antimitotic aphidicolin to the culture at a final concentration of 6 µM.

10. Culture this for several days.

12.3   MALDI-IMS of SCG Neuron

Next, we quickly freeze and dry cultured cells, following the careful removal of the cell culture medium. This fixation process is important to maintain the cellular morphology. Here, for analysis of small molecules, we immediately applied the matrix solution.

1. Prepare the matrix solution before cell fixation. [In this example, suspend well 50 mg DHB (Bruker Daltonics) with 1 ml 70% methanol containing and 20 mM sodium acetate by vigorous vortexing.] Sonicate this solution until the precipitates almost disappear.
2. Wash the culture of SCG with 1 ml 0.1 M phosphate buffer (PB) three times and thoroughly remove the solution.

Note: Residual components of the medium (e.g., proteins, lipids, and salts) degrade samples by the precipitation of the component (see Fig. 12.4).

3. Remove the flexiPERM from the glass slide.
4. After removing the remnant PB thoroughly from the glass slide, freeze it rapidly by placing the glass slide on dry ice or liquid nitrogen.
5. Lyophilize completely under vacuum condition. This process normally takes several hours, though depending on the performance of lyophilizer (result is shown in Fig. 12.4).
6. Spray the matrix solution with the greatest care, to prevent the formation of large beads of solution on the sample. Such beads cause small molecules to disperse,

Fig. 12.4 Photographs of freeze-dried superior cervical ganglion (SCG) neurons, with/without phosphate buffer (*PB*) wash procedure. Adapted from [18] (Reprinted from Yang et al., Proceedings for 7th International Symposium on Atomic Level Characterizations for New Materials and Devices.)

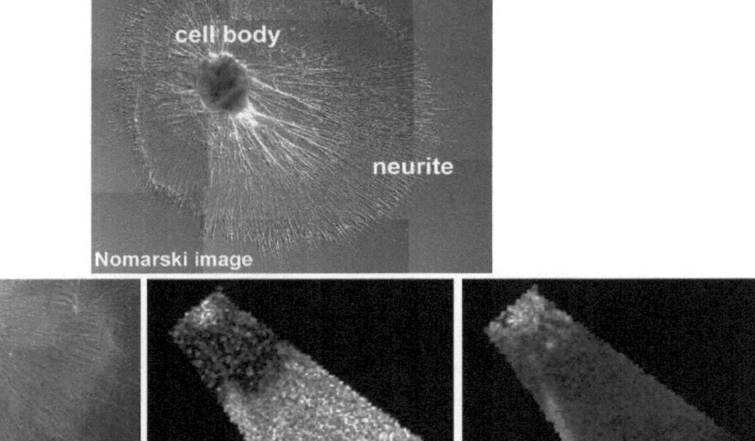

Fig. 12.5 MALDI images of SCG neurons. Nomarski image in culture (*top panel*) and an image after vacuum-freeze drying (*bottom panels*). Optical observation of prepared cells shows that cellular morphology is well maintained after fixation. Furthermore, the IMS result demonstrates heterogeneous molecular distribution within the neurons: ion at *m/z* 734.5 is distributed in neurites and another at *m/z* 826.5 is mainly localized in cell bodies Adapted from [18] (Reprinted from Yang et al., Proceedings for 7th International Symposium on Atomic Level Characterizations for New Materials and Devices.)

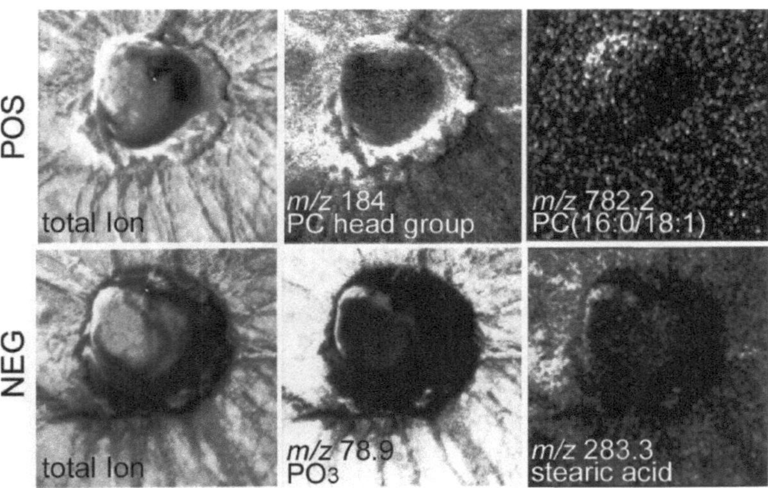

Fig. 12.6 SIMS imaging results for SCG neurons of cell body region. Data were obtained with a SIMS instrument (TRIFT IV, ULVAC-FAI) Adapted from [18] (Reprinted from Yang et al., Proceedings for 7th International Symposium on Atomic Level Characterizations for New Materials and Devices.)

resulting in a loss of signal or the detection of artificial localizations. Detailed spraying methods are described in Chap. 6.

7. The samples are ready to be analyzed by IMS (result is shown in Fig. 12.5).

12.4   Supplemental: SIMS-IMS of SCG Neuron

The sample preparation protocol presented here can be also used for SIMS imaging. For this purpose, follow the aforementioned procedure except matrix application. Figure 12.6 shows an example of SIMS imaging of SCG neurons by a TOF-SIMS (TRIFT IV, ULVAC-PHAI). In the images, relatively small molecules presumably derived from the intact phospholipids (PC head group, phosphate, and stearic acid) were visualized.

References

1. Ostrowski SG, Van Bell CT, Winograd N, Ewing AG (2004) Mass spectrometric imaging of highly curved membranes during *Tetrahymena* mating. Science 305:71–73
2. Rubakhin SS, Greenough WT, Sweedler JV (2003) Spatial profiling with MALDI MS: distribution of neuropeptides within single neurons. Anal Chem 75(20):5374–5380
3. Kruse R, Sweedler JV (2003) Spatial profiling invertebrate ganglia using MALDI MS. J Am Soc Mass Spectrom 14(7):752–759
4. Colliver TL, Brummel CL, Pacholski ML, et al. (1997) Atomic and molecular imaging at the single-cell level with TOF-SIMS. Anal Chem 69(13):2225–2231
5. Ostrowski SG, Kurczy ME, Roddy TP, et al. (2007) Secondary ion MS imaging to relatively quantify cholesterol in the membranes of individual cells from differentially treated populations. Anal Chem 79(10):3554–3560
6. Roddy TP, Cannon DM, Meserole CA, et al. (2002) Imaging of freeze-fractured cells with in situ fluorescence and time-of-flight secondary ion mass spectrometry. Anal Chem 74(16):4011–4019
7. Roddy TP, Cannon DM, Ostrowsk SG, et al. (2002) Identification of cellular sections with imaging mass spectrometry following freeze fracture. Anal Chem 74(16):4020–4026
8. Subhash C, George HM (1995) Imaging ion and molecular transport at subcellular resolution by secondary ion mass spectrometry. Int J Mass Spectrom Ion Process 143:161–176
9. Liu Q, Guo Z, He L (2007) Mass spectrometry imaging of small molecules using desorption/ionization on silicon. Anal Chem 79(10):3535–3541
10. Chaurand P, Schriver KE, Caprioli RM (2007) Instrument design and characterization for high resolution MALDI-MS imaging of tissue sections. J Mass Spectrom 42(4):476–489
11. Altelaar AFM, Taban IM, McDonnell LA, et al. (2007) High-resolution MALDI imaging mass spectrometry allows localization of peptide distributions at cellular length scales in pituitary tissue sections. Int J Mass Spectrom 260(2-3):203–211
12. Shimma S, Sugiura Y, Hayasaka T, et al. (2007) MALDI-based imaging mass spectrometry revealed abnormal distribution of phospholipids in colon cancer liver metastasis. J Chromatogr B 855(1):98–103

13. Sugiura Y, Shimma S, Setou M (2006) Two-step matrix application technique to improve ionization efficiency for matrix-assisted laser desorption/ionization in imaging mass spectrometry. Anal Chem 78(24):8227–8235
14. Altelaar AFM, Klinkert I, Jalink K, et al. (2006) Gold-enhanced biomolecular surface imaging of cells and tissue by SIMS and MALDI mass spectrometry. Anal Chem 78(3):734–742
15. Hillebrandt H, Abdelghani A, Abdelghani-Jacquin C, et al. (2001) Electrical and optical characterization of thrombin-induced permeability of cultured endothelial cell monolayers on semiconductor electrode arrays. Appl Phys A Mater Sci Processing 73(5):539–546
16. Qiu Q, Sayer M, Kawaja M, et al. (1998) Attachment, morphology, and protein expression of rat marrow stromal cells cultured on charged substrate surfaces. J Biomed Mater Res 42(1):117–127
17. Aoki T, Tanino M, Sanui K, et al. (1996) Secretory function of adrenal chromaffin cells cultured on polypyrrole films. Biomaterials 17(20):1971–1974
18. Yang H.J., Zaima N., Sugiura Y. et al. (in press), Imaging of cultured neurons by mass spectrometry., Proceedings for 7th International Symposium on Atomic Level Characteristions for New Materials and Devices

Chapter 13
TLC-Blot-MALDI-IMS

Naoko Goto-Inoue, Takao Taki, and Mitsutoshi Setou

Abstract In this chapter, we demonstrate the technique of TLC-imaging mass spectrometry consisting of thin-layer chromatography (TLC), the TLC-blot method, and imaging mass spectrometry. By the use of this method, we could separate, purify, detect, and identify very faint lipids at the same time. We could detect a lipid at a low-picomole level; this method shows higher sensitivity than the conventional staining method. This method is also suitable for molecular species differential analysis by constructing the ion image of each molecular species with fine resolution (micrometer order). Furthermore, this system is capable of quantitative analysis by the use of TLC. It needs no special equipment or technique, and thus it is easy for everyone to begin using.

13.1 Introduction

Imaging mass spectrometry (IMS) can identify the localization of each material by ionizing not only on sections, but also on membranes. Often, phosphatidylcholine is mainly detected because the ionization efficiency of phosphatidylcholine is higher in a positive-ion mode than other phospholipids [1]. Moreover, its presence might cause the reduction of an unstable chain such as the sialic acid chain of glycolipids by laser irradiation. In this case, it cannot be determined whether the asialo-glycolipids actually exist in the lipid composition or are caused by a reduction of gangliosides.

Therefore, the technique of thin-layer chromatography (TLC)-IMS is adapted to enable the analysis of a small amount of lipid; semiquantitative analysis is also performed.

N. Goto-Inoue and M. Setou (✉)
Department of Molecular Anatomy, Hamamatsu University School of Medicine,
1-20-1 Handayama, Higashi-ku, Hamamatsu, Shizuoka 431-3192, Japan
e-mail: setou@hama-med.ac.jp

T. Taki
Institute of Biomedical Innovation, Otsuka Pharmaceutical Co. Ltd., Kawauchi-cho,
Tokushima 771-0192, Japan

M. Setou (ed.), *Imaging Mass Spectrometry: Protocols for Mass Microscopy*,
DOI 10.1007/978-4-431-09425-8_13, © Springer 2010

In this method, TLC or a transferred polyvinylidene fluoride (PVDF) membrane is directly coupled with mass spectrometry [TLC-matrix-assisted laser desorption/ionization (MALDI) MS or TLC-blot-MALDI MS], which increases the utility of TLC for structural characterization. Recently, there have been some reports regarding TLC-MALDI MS [2–6]; it separates lipids by TLC and also detects, identifies, and quantifies them by staining and analyzing MS directly. In this way, it thus provides both molecular mass and structural information without requiring purification of the sample. We have developed a system that uses the IMS technique (TLC-blot-MALDI IMS); it enables a differential display analysis by utilizing the slight differences in fatty acids; furthermore, ion-trap instruments can perform MS^n analysis to identify detailed structures [7]. It is thought that a comprehensive lipid analysis becomes possible by combining the two techniques of IMS and TLC-Blot-MALDI MS [8].

13.2   TLC Analysis

To undertake TLC analysis, the following materials and reagents are required.

13.2.1   *Materials*

- TLC tank
- Syringe (10 µl)
- Soft pencil (6B)
- High-performance thin-layer chromatography plate (HPTLC-plate) (EM-5631-5; Merck)
- Dryer

13.2.2   *Reagents*

- Chloroform
- Methanol
- Methyl acetate
- Propanol
- 0.25% Potassium chloride

13.2.3   *TLC Method*

TLC is a classical but powerful analytical method for use with all kinds of lipids. Lipid compositions of tissue extracts, quantitative determination, and their chemical identification can be analyzed from their mobility, in combination with chemical reactions, on a TLC plate. The major advantages of this method include the fact that the method and equipment are simple and that most lipids can be separated by choosing the appropriate solvent system. In this chapter, we demonstrate phospholipid analysis by TLC. From our experience, the development solvent methyl acetate/propanol/chloroform/methanol/0.25% potassium chloride (25/25/25/10/9, by volume) is best for the separation of phospholipids:

1. Homogenize the tissue or cell sample with 20 times the volume extraction solvent: chloroform/methanol (2/1, by volume).
2. Before developing the plate, the TLC tank should be saturated with developing solvent.
3. Mark the origins with a soft pencil: 1 cm from the bottom, and widths of 5 mm each.
4. The lipid sample diluted in 2–5 µl (chloroform/methanol=2/1, by volume) is carefully and evenly applied to these pencil marks.
5. After applying the sample, the HPTLC-plate is heated with a dryer for 1 min and then developed in a TLC tank with the developing solvent.
6. When the solvent front reaches approximately 7–8 cm from the bottom, the plate is taken and the solvent on the plate is removed with a dryer.

13.3 Primuline Staining

To undertake primuline staining, the following materials and reagents are required.

13.3.1 Materials

- Spray bottle
- UV light (315 nm)
- Soft red pencil
- Safety glass

13.3.2 Reagents

- Stock solution: 100 mg primuline is dissolved in 100 ml water
- Spray reagent: 1 ml of the stock solution is diluted in 100 ml of a mixture of acetone/water (4/1, by volume)

13.3.3 Primuline Staining Method

Primuline staining is the most sensitive and highest quantitative staining method for detecting all lipids. It is a staining method that has no chemical influence on MS analysis, because almost all primuline is washed after transfer to the PVDF membrane. Further, it can be used together with the ninhydrin reagent in the detection of amino group-containing lipids. One can also choose to detect the phospholipid on the TLC plate with other reagents, such as the Dittmer reagent, although it is unsuitable for TLC-blot, because the reagent includes strong sulfuric acid.

The steps are as follows:

1. Prepare the HPTLC plate after TLC development.
2. The HPTLC-plate is sprayed uniformly with primuline reagent.
3. After drying the HPTLC-plate, the separated lipid bands are detected under UV light (315 nm) in a dark room. When you use UV light, safety glasses should be worn to protect the eyes.
4. The position and the shape of the detected bands are marked with a soft red pencil.

13.4   TLC-Blot

To undertake TLC-blot, we use the following materials and reagents.

13.4.1   Materials

- TLC thermal blotter (AC-5970; ATTO)
- Glass fiber filter sheets (AC-5972; ATTO)
- Teflon membrane (PIFE membrane; ATTO)
- PVDF membrane (Clear Blot membrane-P; ATTO)

13.4.2   Reagents

- Isopropanol
- 0.2% $CaCl_2$
- Methanol

13.4.3   TLC-Blot Method

Although it is possible to analyze a HPTLC-plate directly, a transferred PVDF membrane shows a higher sensitivity and resolution than a HPTLC-plate. The transfer procedure is simple and rapid and shows high efficiency [9–11]. Established lipid detection procedures – including chemical reagents, immunostaining, enzyme assays, and binding studies – are available for blotted membranes. Furthermore, the transferred lipids on the membrane are stable. For example, phospholipids on the transferred PVDF membrane can be stored for several months in a cool dark room and analyzed by TLC-blot-MALDI MS after storage, according to the following steps:

1. Prepare the stained HPTLC-plate where the sample is marked with soft red pencil.
2. The plate is dipped in a blotting solvent: isopropanol/0.2% $CaCl_2$/methanol (40/20/7, by volume) for 10 s.
3. Take the plate and cover it immediately with a PVDF membrane.
4. A Teflon membrane and a glass filter paper are then mounted.
5. This assembly is transferred to the TLC thermal blotter, which is heated to 180°C in advance. If this apparatus is not available, a 180°C heated iron is available for TLC-blot.
6. Press the assembly for 30 s and remove the PVDF membrane.
7. Lipids separated on the HPTLC-plate are transferred with the red pencil marker. The lipids will be found along with the blotted pencil dye.

13.5 Preparation for MS Analysis

To undertake MS analysis, the following materials and reagents are required.

13.5.1 *Materials*

- MALDI-QIT-TOF-MS (AXIMA-QIT; Shimadzu)
- Chemical inkjet printer (ChIP-1000; Shimadzu)
- Electrified double-adhesive tape (241-08728-92; Shimadzu)
- MALDI target plate (241-08727-91; Shimadzu)

13.5.2 *Reagents*

- 2,5-Dihydroxybenzoic acid (DHB, 201346; Bruker Daltonics)
- Methanol
- Trifluoroacetic acid (TFA)

13.5.3 *Method of Preparing for MS Analysis*

The following details the method of preparing a sample for MS analysis, i.e., applying a matrix solution to the PVDF membrane. Refer to Chap. 8 for practical imaging processing that uses the MALDI target plate where the matrix is applied:

1. Prepare the PVDF membrane transferred by TLC-blot.
2. Be careful that samples are transferred to the PVDF membrane so as not to face the HPTLC-plate but the reverse side. Attach the PVDF membrane to a MALDI target plate with electrified double-adhesive tape to reduce the charge-up of the plate.
3. The distribution of each band on the plate is quantified by using ChIP-1000, which forwards it on to AXIMA-QIT.
4. Apply the matrix solution: 50 mg/ml DHB in methanol/0.1% TFA solution (1/1, by volume) 5 µl/band, twice. The crystal is clearly seen, and it discolors to yellow on the PVDF membrane. Afterwards, dry the membrane well.
5. Because applying the matrix to the membrane can lead to a rough surface, the matrix layer should be evenly flattened with a thick wiper (5–10 s).
6. The target plate with the PVDF membrane is set into AXIMA-QIT, whereupon measurement begins. When the signal intensity appears weak, it is best to apply the matrix solution again.

13.6 Imaging Analysis

This section discusses the example of sphingomyelin analysis, which involves one of the standard phospholipids, measured by the before mentioned TLC-blot-MALDI IMS methodology. A workflow of this method is shown in Fig. 13.1. All processes are very easy and efficient, compared to conventional multicolumn chromatographic processes; the entire process from TLC analysis to MS analysis can be performed in approximately half a day.

At first, TLC analysis was performed according to the common procedure and staining with primuline reagent (Fig. 13.2). The sphingomyelin developed on the right lane. As a result, a single broad band could be detected; however, it is thought that two or more molecular species were included in this band.

Next, the result of the TLC-blot-MALDI IMS analysis of the same sample is shown (Fig. 13.3). Although all detections are consolidated in one band when using the staining method (see Fig. 13.2), there are plural molecular species and each molecular species forms two or more layers. Here we show the result of three kinds of sphingomyelin molecular species. Shorter-chain fatty acids containing sphingomyelin moved slower than those containing longer chains with highly unsaturated fatty acids. Applying TLC-blot-MALDI IMS would visualize sphingomyelin and distinguish the minute moiety patterns that are developed in TLC, based on information regarding their mass sizes or hydrophobicity; they enable us to discriminate the molecular species in greater detail than the previous staining procedure.

Moreover, it is possible to obtain the detailed structural characterization of each molecular species by MS/MS analysis. Therefore, this method is suitable for the analysis of molecular species compositions or separate structural analyses, and it might prove to be a powerful tool for lipidomics.

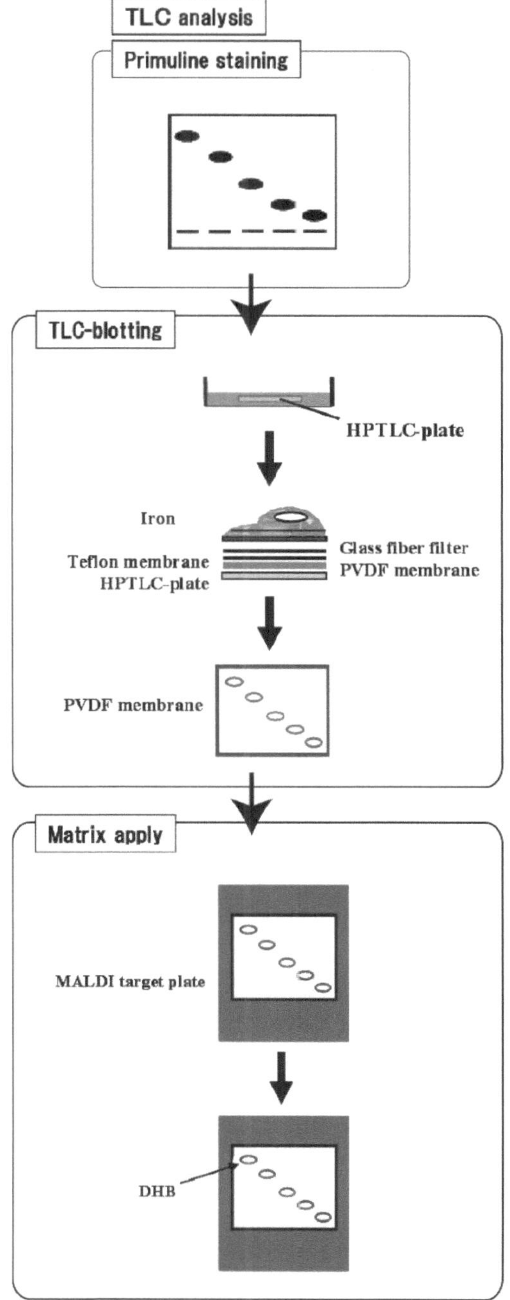

Fig. 13.1 Work flow of thin-layer chromatography (TLC)-blot-imaging mass spectrometry (IMS). *PVDF*; polyvinylidene fluoride, *MALDI*; matrix-assisted laser desorption/ionization

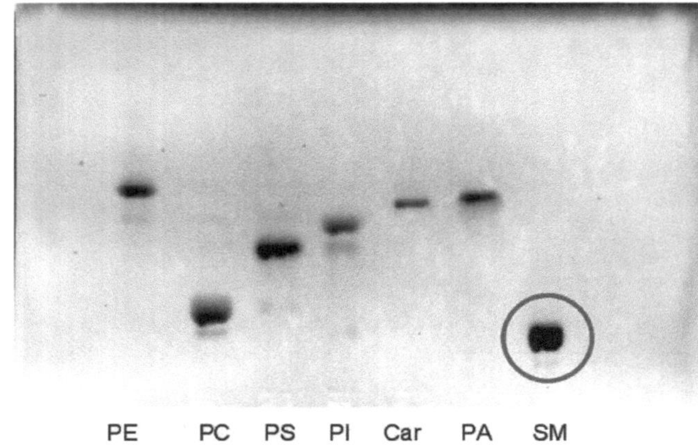

Fig. 13.2 A result of primuline staining of standard phospholipid samples (*PE*; phosphatidyle-thanolamine, *PC*; phosphatidylcholine, *PS*; phosphatidylserine, *PI*; phosphatidylinositol, *Car*; cardiolipin, *PA*; phosphatidylamine, *SM*; sphingomyelin)

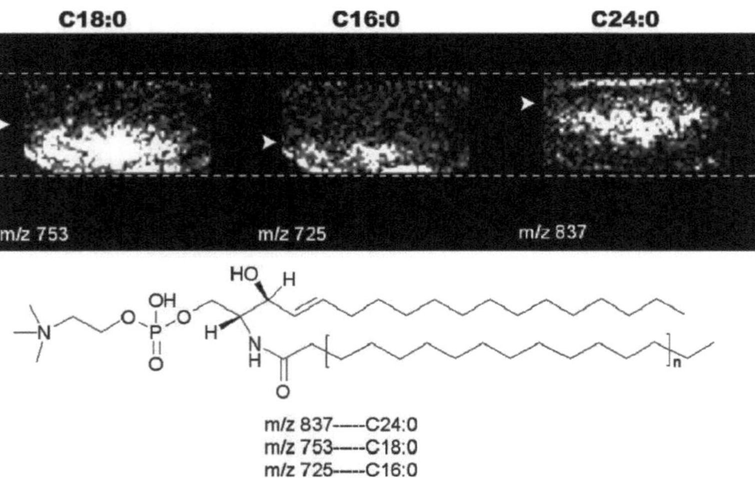

Fig. 13.3 The result of TLC-blot-MALDI IMS and structural information of each molecular species of sphingomyelin

References

1. Petkovic M, Schiller J, Müller M, et al. (2001) Detection of individual phospholipids in lipid mixtures by matrix-assisted laser desorption/ionization time-of-flight mass spectrometry: phosphatidylcholine prevents the detection of further species. Anal Biochem 289(2):202–216

2. Dreisewerd K, Müthing J, Rohlfing A, et al. (2005) Analysis of gangliosides directly from thin-layer chromatography plates by infrared matrix-assisted laser desorption/ionization orthogonal time-of-flight mass spectrometry with a glycerol matrix. Anal Chem 77(13):4098–4107

3. Ivleva VB, Elkin Y, Budnik BA, et al. (2004) Coupling thin-layer chromatography with vibrational cooling matrix-assisted laser desorption/ionization Fourier transform mass spectrometry for the analysis of ganglioside mixtures. Anal Chem 76(21):6484–6491

4. Ivleva VB, Sapp L, O'Connor PB, et al. (2005) Ganglioside analysis by thin-layer chromatography matrix-assisted laser desorption/ionization orthogonal time-of-flight mass spectrometry. J Am Soc Mass Spectrom 16(9):1552–1560

5. Nakamura K, Suzuki Y, Goto-Inoue N, et al. (2006) Structural characterization of neutral glycosphingolipids by thin-layer chromatography coupled to matrix-assisted laser desorption/ionization quadrupole ion trap time-of-flight MS/MS. Anal Chem 78(16):5736–5743

6. O'Connor PB, Budnik B, Ivleva VB, et al. (2004) A high pressure matrix-assisted laser desorption ion source for Fourier transform mass spectrometry designed to accommodate large targets with diverse surfaces. J Am Soc Mass Spectrom 15(1):128–132

7. Goto-Inoue N, Hayasaka T, Sugiura Y, et al. (2008) High-sensitivity analysis of glycosphingolipids by matrix-assisted laser desorption/ionization quadrupole ion trap time-of-flight imaging mass spectrometry on transfer membranes. J Chromatogr B Anal Technol Biomed Life Sci 870(1):74–83

8. Goto-Inoue N, Hayasaka T, Taki T, et al. (2009) A new lipidomics approach by thin-layer chromatography-blot-matrix-assisted laser desorption/ionization imaging mass spectrometry for analyzing detailed patterns of phospholipid molecular species. J Chromatogr A 1216: 7096–7101

9. Guittard J, Hronowski X, Costello CE (1999) Direct matrix-assisted laser desorption/ionization mass spectrometric analysis of glycosphingolipids on thin layer chromatographic plates and transfer membranes. Rapid Commun Mass Spectrom 13(18):1838–1849

10. Ishikawa D, Taki T (2000) Thin-layer chromatography immunostaining. Methods Enzymol 312:145–157

11. Taki T, Ishikawa D, Handa S, et al. (1995) Direct mass spectrometric analysis of glycosphingolipid transferred to a polyvinylidene difluoride membrane by thin-layer chromatography blotting. Anal Biochem 225:24–27

Part VII
IMS Applications Provided by Manufacturers

Chapter 14
Applied Biosystems

Tetsuo Kokaji

Abstract The oversampling method succeeded in improving the resolution of imaging mass spectrometry (IMS) without any changes to the laser optics of their instrument. In addition, other useful tools have been also developed: Dynamic pixel imaging (patent pending) is a technique that can allow longer mass spectrometry (MS) analysis time over a defined area during a IMS experiment and it can also be used as a multiplexing tool such as dependent scan when MS to MS/MS data are simultaneously acquired. Here we report three analytical examples using these new measuring technologies.

14.1 Introduction

Matrix-assisted laser desorption/ionization mass spectrometry (MALDI-MS) imaging is a highly valuable tool in determining the localization/distribution of biological components and drugs administered directly from the surface of a tissue sample. Imaging mass spectrometry (IMS) has a number of advantages over traditional imaging methods such as autoradiography, or immunochemical imaging, including faster data acquisition, the ability to analyze multiple compounds, and a simpler workflow (i.e., no compound labeling is required). Furthermore, compound-specific structural information may be obtained by performing MS/MS-based MALDI-IMS.

Recent advances in IMS techniques have included the development of a sprayer for matrix application; the parallel exhibition of mass spectral images at multiple m/z values in data processing; overlapping with an optical image; and the generation of clearer images by correcting signals from the matrix using a normalization process. In addition, the spectral data obtained can be statistically analyzed (Fig. 14.1).

T. Kokaji (⊠)
Applied Biosystems Japan, 4-5-4 Hacchobori, Chuo-ku, Tokyo 104-0032, Japan
e-mail: Tetsuo.Kokaji@lifetech.com

M. Setou (ed.), *Imaging Mass Spectrometry: Protocols for Mass Microscopy*,
DOI 10.1007/978-4-431-09425-8_14, © Springer 2010

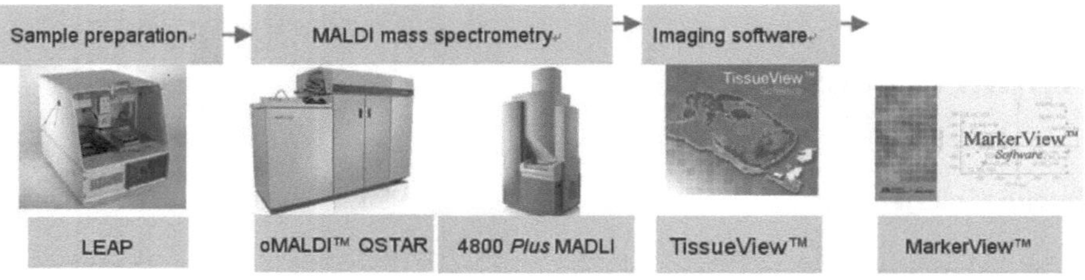

Fig. 14.1 Overview of imaging mass spectrometry (IMS) system provided by Applied Biosystems

With regard to measuring technology, an oversampling method that allows greater IMS resolution has been reported by Jurchen et al. Their method succeeded in improving the resolution of IMS, without requiring any changes to the laser optics of their instrument. In addition, other useful tools have been also developed. Dynamic pixel imaging (DPI; patent pending) is a technique that can allow a longer MS analysis time over a defined area during an IMS experiment; it can also be used as a multiplexing tool such as dependent scan when MS to MS/MS data are simultaneously acquired. Here we report three analytical examples using these new measuring technologies.

14.2   Imaging of Rat Brain Sections on the 4800 MALDI-TOF/TOF Analyzer

As mentioned previously, MALDI-MS imaging is a technique that allows for the analysis of biomolecules directly on the tissue samples [1–4]. The technique involves mounting a thinly sliced tissue section onto a sample plate and applying MALDI matrix directly to the top of the tissue. A two-dimensional (2D) array of MS spectra is obtained for the entire area scanned; from this, an ion intensity map is produced for any mass that is detected within the scanned area. Tissue samples can range from a small section of tissue to entire organs or even the entire body of an animal.

MALDI-MS offers a number of advantages over other imaging technologies:

- The sample is analyzed directly
- No extraction of compounds is required
- No labeling of the sample is needed

In the 4800 MALDI-TOF/TOF Analyzer, the Opti-Beam on-axis laser system provides unsurpassed sensitivity, with a fast 200-Hz firing rate. Its large target plate permits the imaging of larger tissue sections. Combined with the BioMAP software – which is, as discussed previously (see Chap. 9), a potentially powerful tool for imaging processing in itself – an ideal MS imaging system can be constructed.

In this measurement, we found the presence of a protein localized in rat brain by using the 4800 MALDI-TOF/TOF Analyzer.

14.2.1 Experimental Methods

Targeted tissues were collected from healthy control rats, frozen, and stored at −80°C. Brain tissue sections were cut on a Leica cryostat maintained at −17°C and thaw-mounted on stainless steel MALDI target plates. The tissues on the plate were washed with 90% ethanol, treated with 10 mg/ml α-cyano-4-hydroxycinnamic acid (α-CHCA) solution containing 10% water, and spray-coated with 20 mg/ml α-CHCA (dissolved in 50:50 acetonitrile: 0.2% trifluoroacetic acid in water). All MS imaging analyses on the samples pretreated in the aforementioned procedure were performed by the 4800 MALDI-TOF/TOF Analyzer.

As for spectrum measurement, the standard linear and reflector modes were used. Data were collected using a MS imaging-acquisition tool available at http://www.maldi-msi.org/. The pixel size was set to 200 μm × 200 μm, and 50 laser shots were summed per pixel.

Linear spectra were acquired over the mass range of 5–75 kDa, and reflector data were acquired over the range of 1–4 kDa. All imaging processing was performed using BioMAP software (available at http://www.maldi-msi.org/).

14.2.2 Results

Figure 14.2 is an image from signals at different *m/z* values obtained in reflector mode; it shows the corresponding ion intensity profiles for the spectra selected. This result indicates the dramatic differences in the localization of biological components selected across the rat brain section.

An image based on spectra obtained in linear mode is shown in Fig. 14.3. Data were collected by scanning over the mass range of 5–75 kDa. As shown in the reflector mode data, specific components are seen to be localized in specific regions of the brain sections. By overlaying the image on an optical image (i.e., a picture of the tissue), a correlation between the signal intensities and the optical image can be observed.

Next, the identification of specific components of interest from the tissue image data was attempted by MS/MS analysis. The peak at *m/z* 2,510 that appears to be concentrated in two different areas of the sample (see Fig. 14.2, left panel) was selected for MS/MS. The MS/MS spectrum was acquired across the entire sample, and the image for the ion transition from *m/z* 2,510 to *m/z* 1,378 was plotted using BioMAP software (Fig. 14.4, inset). The resultant image had good agreement with the MS spectrum image.

At this point, the MS/MS spectra from a region where the components were concentrated (indicated by white dashed line) were summed to a single MS/MS spectrum.

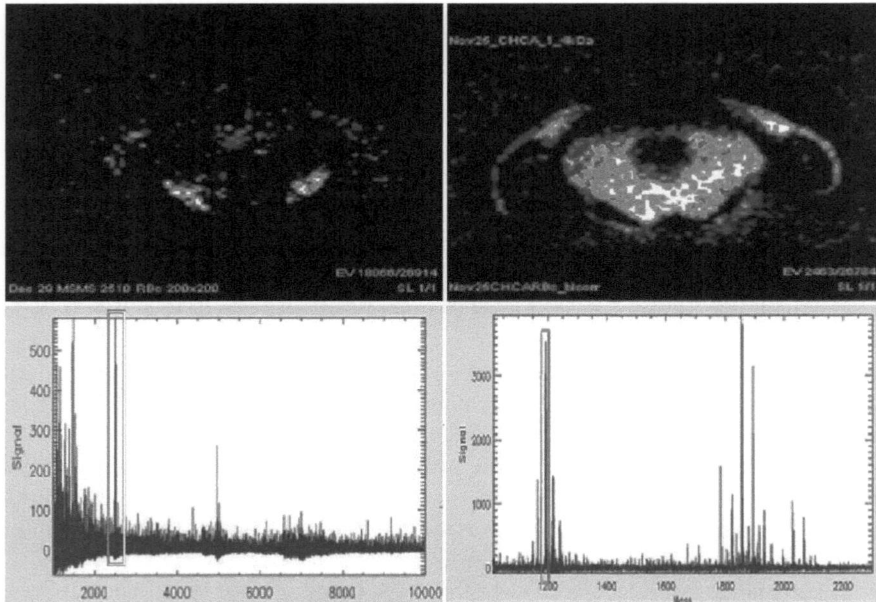

Fig. 14.2 Reflector-mode IMS spectra and corresponding images from coronal rat brain sections produced by BioMAP software. *Right panel* represents the distribution of the peak at 1,190 Da across the tissue section; *left panel* shows the image of the peak at 2,510 Da. The intensity of the selected mass at any particular point on the image is represented by color intensity (*red* indicates the most intense)

This spectrum was searched against the NCBI nr database using the Mascot search engine. The Mascot search results produced an MS/MS ion probability score of 67 for the peptide sequence and a confident match to the protein prodynorphin (see Fig. 14.4). Prodynorphin is an opioid precursor protein of a bioactive peptide that is consistent with opioid activity in the rat brain. Specific in vivo processing of prodynophin is responsible for the production of a number of peptides as well as many neuropeptides.

14.2.3   Conclusion

Imaging by MALDI-MS represents a new methodology for biomarker discovery, as derived directly from tissue samples. The distribution of known or potentially new biomarkers may be profiled across a tissue section, without the need for extraction or the labeling of samples.

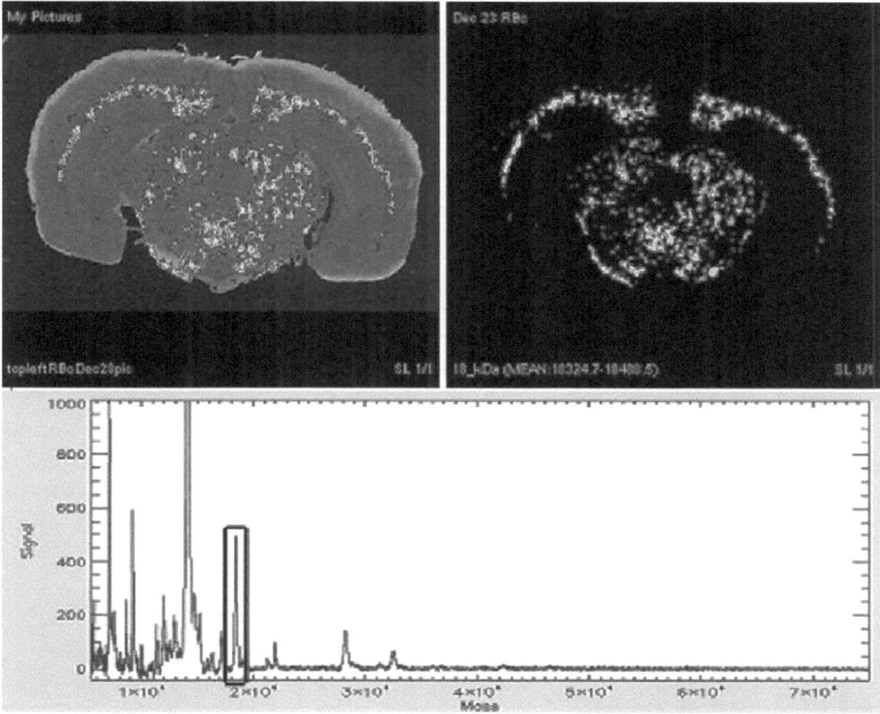

Fig. 14.3 Linear-mode spectra and image corresponding to the peak at 18,375 Da. The BioMAP software allows IMS images to be overlaid on optical images of the tissue sample, making direct correlations between IMS profiles and optical features possible

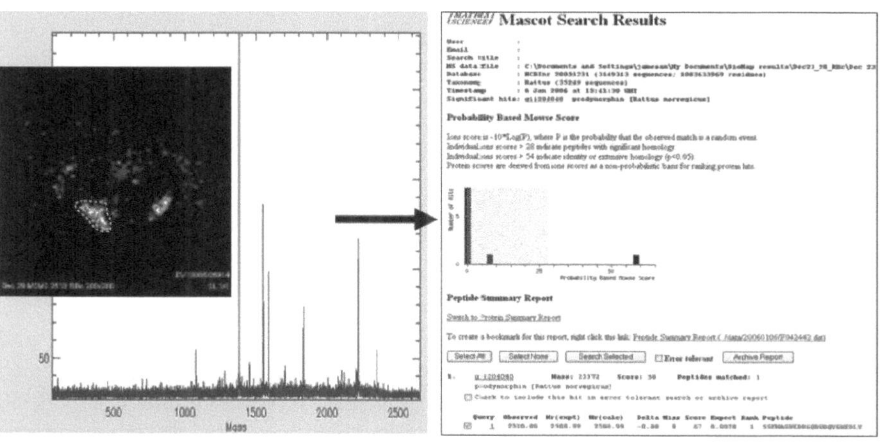

Fig. 14.4 On-tissue feature identification by MS/MS and database searching on the 4800 MALDI-TOF/TOF Analyzer. The MS/MS spectra over a particular region of interest may be summed together, and the resultant spectrum can be searched against the NCBI nr database using the Mascot search engine

14.3 Multiplexing MALDI-MS Imaging Using DPI

It is relatively easy to produce imaging data of multiple parent compounds in a time-of flight mass spectrometry (TOF-MS) experiment. However, this type of analysis lacks compound specificity and can lead to erroneous results because of the reduction of signal intensities by ion suppression effects.

Here we developed a new imaging technique for the acquisition of multiple MS/MS and TOF-MS data that is called, as previously introduced, DPI. It is a technique that can allow a longer MS analysis time over a defined area during MALDI-MS imaging; it can also be used as a multiplexing tool when MS to MS/MS data are simultaneously acquired.

DPI was used in the simultaneous IMS analysis of the drug Diazepam and its two metabolites – namely, Temazepam and Desmethyl Diazepam – in rat liver tissue sections. In conjunction with DPI, a new processing technique is discussed for the normalization of resultant IMS data.

14.3.1 Materials and Methods

14.3.1.1 Instrumentation

An Applied Biosystems/MDS Sciex QSTAR Elite QqTOF-MS with an oMALDI 2 ion source, equipped with an Nd:YAG laser, was used. The laser energy was optimized for CHCA. TOF-MS and MS/MS experiments were carried out simultaneously and analyzed separately, as needed.

14.3.1.2 Software

The Analyst QS 2.0 and oMALDI Server 5.0 software packages were used to obtain mass spectra, and oMALDI Server 5.0 and BioMAP were used to process data.

14.3.1.3 Sample Preparation

A male rat was given a single oral dose of Diazepam (10 mg/kg). The liver and brain were removed 1 h postdose and flash-frozen on dry ice. Tissues were sliced to a thickness of 20 µm using a Leica Microsystems (Wetzlar, Germany) CM 3050 cryomicrotome maintained at −25°C; they were then thaw-mounted on a stainless steel MALDI plate (Fig. 14.5).

Multiple coats of CHCA matrix solution (20 mg/ml) were applied to plated tissue samples using a TLC sprayer (Sigma). (See Fig. 14.5 for a full description of the IMS workflow.)

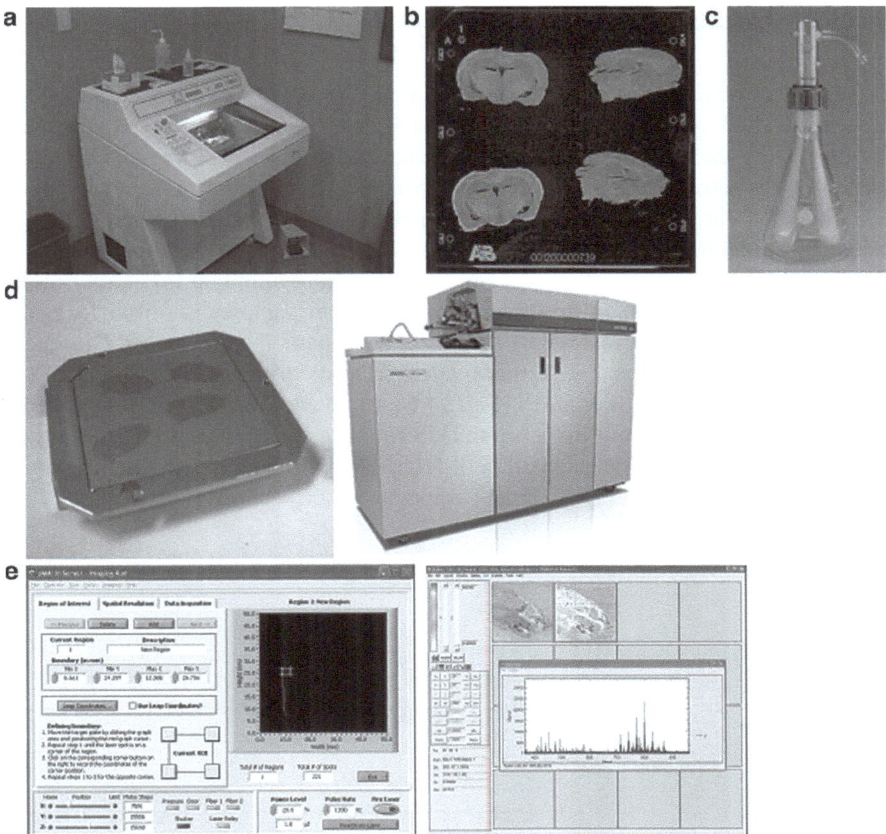

Fig. 14.5 Workflow for matrix-assisted laser desorption/ionization-mass spectrometry (MALDI-MS) using the QSTAR Elite system: slice tissue sections (**a**), mount sample to plate (**b**), spray-coat with matrix (**c**), load sample plate into QSTAR Elite system (**d**), and perform data acquisition using oMALDI server 5.0 software and processing with BioMAP software (**e**)

14.3.1.4 Acquisition Parameters

The mass range of product ion scans were set for the following compounds, as follows: Diazepam, 285 (precursor) 145–250 Da; Temazepam, 301 (precursor) 145–250 Da; Desmethyl Diazepam, 271 (precursor) 145–250 Da. A TOF-MS (parent ion) experiment was acquired at 300–1,200 Da. The imaging resolution was set to 300 mm × 300 mm for DPI.

14.3.2   Data Acquisition

MALDI-MS involves the analysis of a 2D area. The area to be scanned is divided into pixels, and an MS spectrum is acquired at every pixel. An image is generated by displaying the intensity of a detected peak at each pixel over the area scanned. When DPI is activated, there is more time available to acquire data from each pixel within an IMS area. For example, looped methods can be designed where different acquisition methods are run consecutively. When a looped acquisition method is combined with DPI, multiple images can be acquired within one IMS experiment, to a maximum of eight. In this work, using rat liver samples removed from the animal killed 1 h after oral dosage of the drug Diazepam, the unchanged Diazepam and two of its metabolites (i.e., Temazepam and Desmethyl Diazepam) produced in the liver were analyzed by MS/MS (Fig. 14.6a–c). In addition, a TOF-MS profiling experiment was carried out in the same IMS run (Fig. 14.6d), resulting in four different scans. Typically, the acquisition of four MALDI imaging experiments would require four separate runs and, perhaps, separate tissue sections.

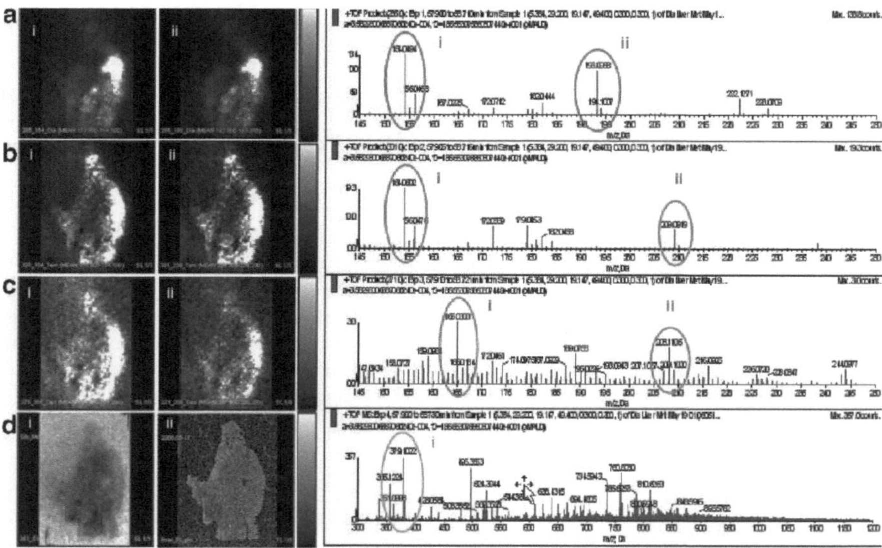

Fig. 14.6 Imaging multiple compound fragments for structural confirmation. There is sufficient information in the spectral data from each MS/MS IMS experiment to produce images of multiple fragments from a given parent; this can be useful for absolute confirmation, to provide compound identity. The following images are displayed with accompanying experimental spectra: MS/MS images of the $285 \rightarrow 154$ (i) and $285 \rightarrow 193$ (ii) transitions of Diazepam (**a**); MS/MS images of the $301 \rightarrow 154$ (i) and $301 \rightarrow 209$ (ii) transitions of Temazepam (**b**); MS/MS images of the $371 \rightarrow 165$ (i) and $371 \rightarrow 208$ (ii) transitions of Desmethyl Diazepam (**c**); and TOF-MS image of the second isotope of the CHCA matrix cluster at *m/z* 379 (i) and optical image of the rat liver tissue section (ii) (**d**). Each of the fragment images produced from a unique precursor had the exact same distribution; there were, however, differences in overall intensity

14.3.3   *Results*

When the images for Diazepam, Temazepam, and Desmethyl Diazepam were set to the same intensity threshold, a quantitatively proportional image for the three compounds was observed. In the case of Diazepam and Temazepam (Fig. 14.7a, b), the signal is so intense that the image is white. On the other hand, Desmethyl Diazepam is much less intense (Fig. 14.7c) and is shown in green. Considering that the animal was killed 1 h after dosing, the results are also consistent with the observation that

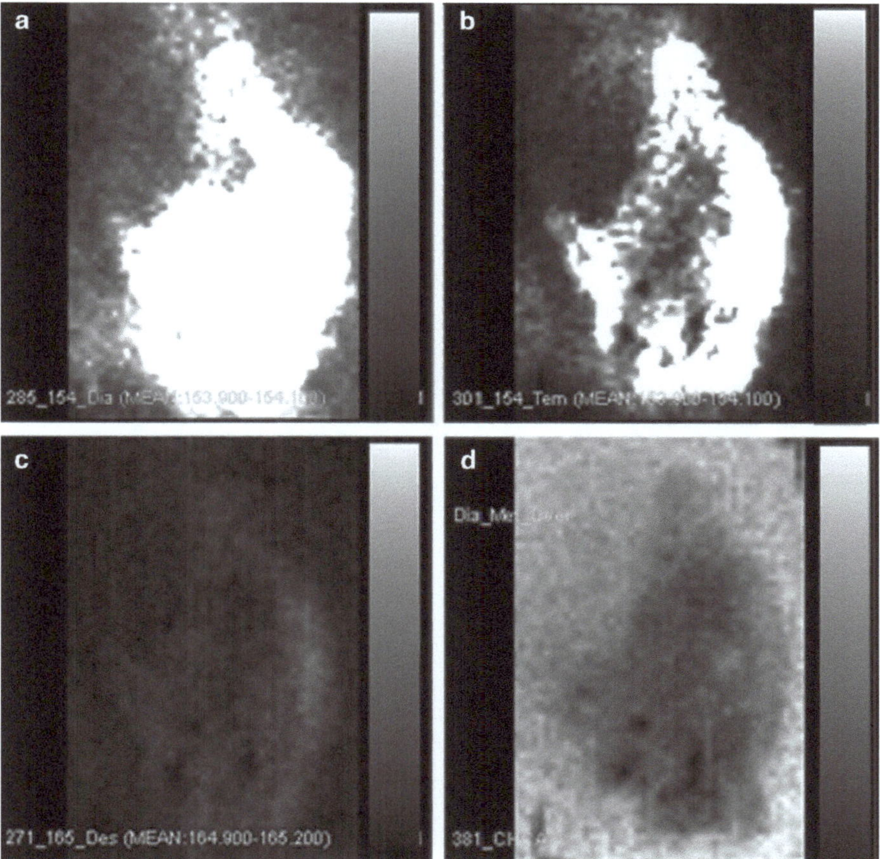

Fig. 14.7 Relative intensities for MS images of Diazepam and its metabolites acquired from rat liver sections. (**a**) MS/MS image of the $285 \rightarrow 154$ transition of Diazepam. (**b**) MS/MS image of the $301 \rightarrow 154$ transition of Temazepam. (**c**) MS/MS image of the $371 \rightarrow 165$ transition of Desmethyl Diazepam. (**d**) TOF-MS image of a cyano-4-hydroxy cinnamic acid (CHCA) matrix cluster at m/z 381 (this is the second C-13 isotope of the 379 cluster). All image experiments were acquired simultaneously using dynamic pixel imaging. The intensity thresholds for (**a**)–(**c**) have been set to the same level to compare relative concentration

the parent drug Diazepam would still be the most abundant species. The metabolic transformations that occurred in the liver to produce Temazepam and Desmethyl Diazepam would have been occurred at different rates.

When the drug/metabolite ion intensities are optimized for image quality, differences in their distribution profiles within the tissue became apparent (Fig. 14.8).

When MALDI-MS is performed directly from the surface of a tissue sample – especially where limited sample cleanup has been performed – there are likely ion suppression effects caused by thousands of species competing for ionization. Different parts of a tissue may have different salt concentrations, resulting in an inaccurate distribution of molecules because of competition for ionization.

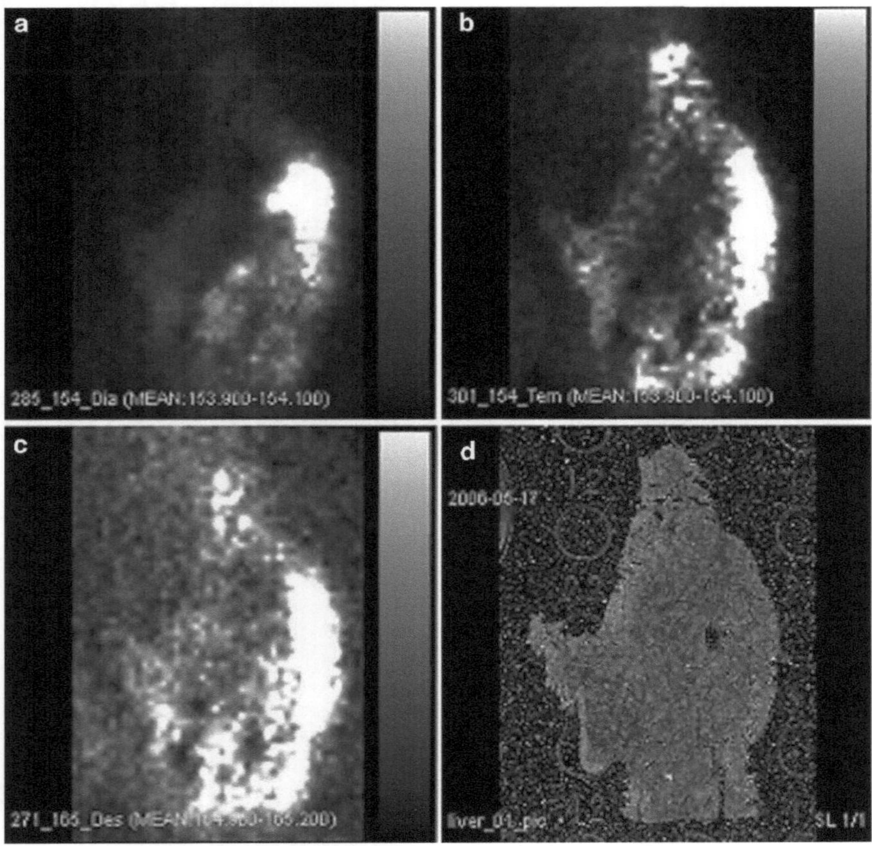

Fig. 14.8 MS images of Diazepam and metabolites acquired from a rat liver section, optimized for each analyte. (**a**) MS/MS image of the 285 → 154 transition of Diazepam (*red*). (**b**) MS/MS image of the 301 → 154 transition of Temazepam (*blue*). (**c**) MS/MS image of the 371 → 165 transition of Desmethyl Diazepam (*green*). (**d**) Optical image of the liver tissue section. All experiments were acquired simultaneously using dynamic pixel imaging. Each IMS analyte was optimized for image distribution and appearance

The normalization of resultant images, using a control compound, should correct for any variation in ionization efficiency for an analyte of interest. To normalize the data, the MS image of an analyte is divided by the CHCA image. In this case, the CHCA image is a control compound, and it can be used to correct for any variations in ion current over the area analyzed.

A final set of images was produced when the IMS data were normalized against the image of a common CHCA cluster (m/z 379.09, $[2\,M+H]^+$) (Fig. 14.9).

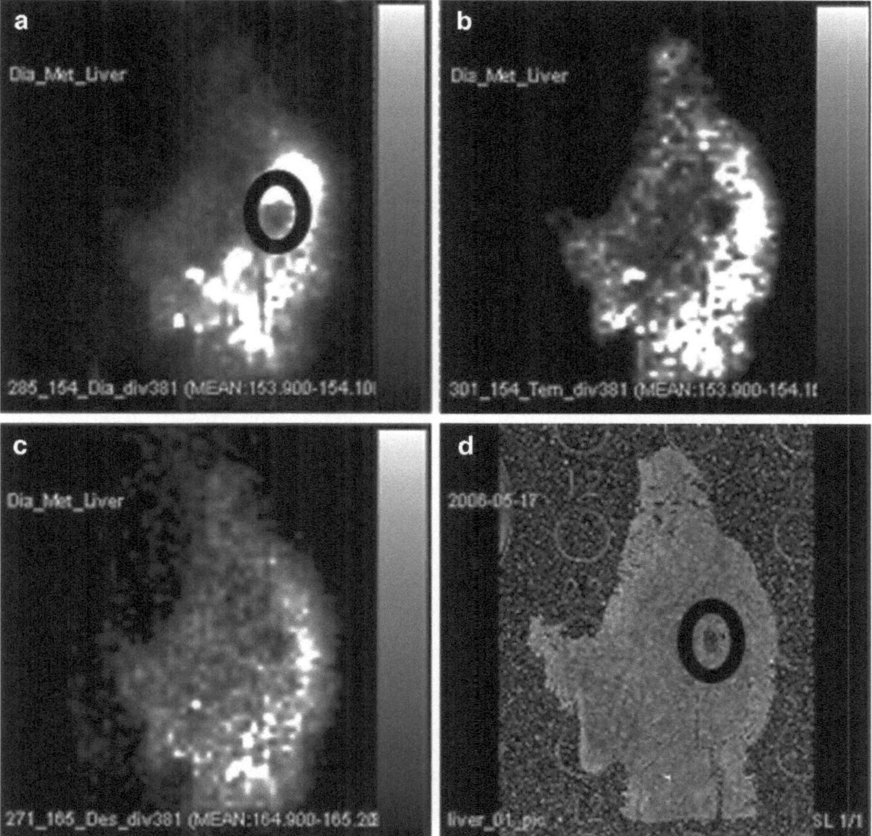

Fig. 14.9 Norm alized rat liver section: MS images of Diazepam and its metabolites. (**a**) MS/MS image of the 285 → 154 transition of Diazepam. (**b**) MS/MS image of the 301 → 154 transition of Temazepam. (**c**) MS/MS image of the 371 → 165 transition of Desmethyl Diazepam. (**d**) Optical image of the liver section before matrix application. Images were normalized by dividing raw data images (**a**)–(**c**) by the TOF-MS image of CHCA at 381 (**d**)

14.3.4   Summary

DPI allows for multiplexing of multiple IMS experiments from a single imaging run. It is possible to image a dosed drug from a tissue of interest, along with known metabolites. Differences in drug and metabolite distribution were observed. Image normalization produced clearer images that better reflected a distribution of each compound across the tissue.

14.4   High-Resolution MALDI-MS Imaging

Obtaining better spatial resolution can be critical to obtaining anatomical findings from the regions of interest in tissue samples. However, image resolution in MALDI-MS has traditionally been limited by laser spot size. Recently, Jurchen et al. reported an oversampling method that allows greater IMS resolution, without any changes to the laser optics of the instrument [5]. In this section, we report the application of this method to a real sample.

The brain of a rat orally given Diazepam (10 mg/kg) was used. The laser spot size on the QSTAR Elite system with a high repetition Nd:YAG laser was approximately 120 μm$\times$200 μm (oval shape). Using the oversampling method, we were able to acquire a high-resolution MALDI tissue image where the pixel size was 60 μm$\times$100 μm, equivalent to one quarter of the area of the laser spot. In the oversampling method, the raster increment of the sample stage movement is smaller than the diameter of the laser beam, with complete MALDI ablation at every pixel. Because the spectral signal is dependent upon the matrix used, only the fresh region of sample produces an ion current with each overlapping area of ablation. This effectively increases the resolution in an IMS experiment, without altering the laser spot size. As such, the image resolution can be improved to the point that specific physiological features of the brain can clearly defined therein.

14.4.1   Materials and Methods

14.4.1.1   Instrument

An Applied Biosystems/MDS Sciex QSTAR Elite QqTOF-MS equipped with an oMALDI 2 ion source with an Nd:YAG laser was used. The laser energy was optimized for CHCA. TOF-MS and MS/MS were simultaneously measured and, as needed, a single measurement was carried out.

14.4.1.2   Software

The Analyst QS 2.0 and oMALDI Server 5.0 software packages were used to obtain mass spectra, and oMALDI Server 5.0 and BioMAP were used to process data.

14.4.1.3 Sample Preparation

A male rat was given a single oral dose of Diazepam (10 mg/kg). The liver and brain were removed 1 h postdose and flash-frozen on dry ice. Tissues were sliced to a thickness of 20 μm using a Leica Microsystems CM 3050 cryomicrotome maintained at −25°C and thaw-mounted to a stainless steel MALDI plate (see Fig. 14.5).

Multiple coats of CHCA matrix solution (20 mg/ml) were applied to plated tissue samples using a TLC sprayer (Sigma).

A pure standard of Diazepam was mixed with CHCA (6 mg/ml in 50% acetonitrile + 50% water) and deposited onto a MALDI plate. The Diazepam standard was used to optimize the MS/MS condition (Fig. 14.10).

14.4.1.4 Method

The TOF-MS range was 300–1,200 Da, whereas the product ion mass range for Diazepam 285 (the precursor) was 145–250 Da (collected). Finally, the Nd:YAG laser high repetition rate was 500 Hz, 4.0 mJ.

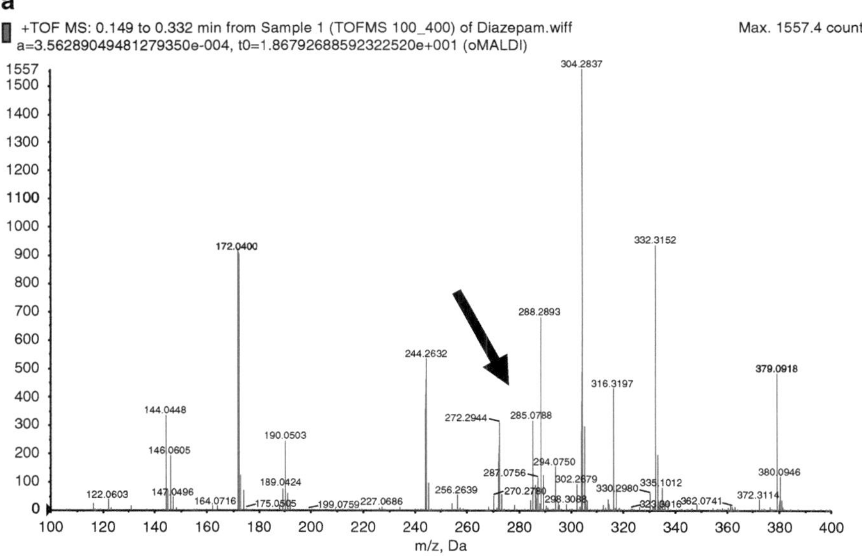

Fig. 14.10 MALDI-TOF-MS and MS/MS spectra obtained for Diazepam. All spectra were acquired from a standard solution of Diazepam. First, a two-point external calibration was performed using the 190 and 379 matrix peaks. A TOF-MS spectrum of Diazepam was acquired (**a**). This spectrum illustrates how much interference can come from MALDI matrix in a mass range <1,000 Da. (**b**) An enlargement of the region around the Diazepam parent; mass accuracy was measured to be 0.35 ppm. (**c**) An MS/MS spectrum was then acquired on the Diazepam standard

(continued)

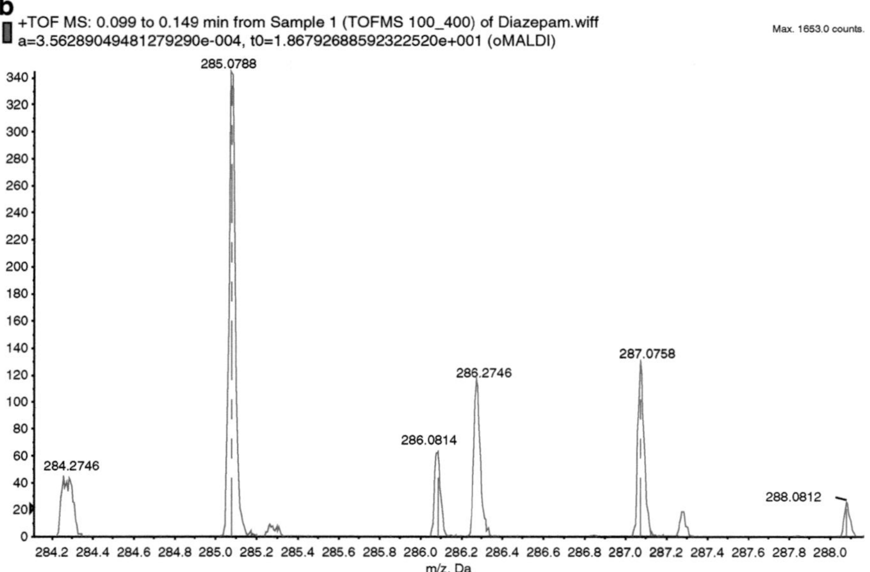

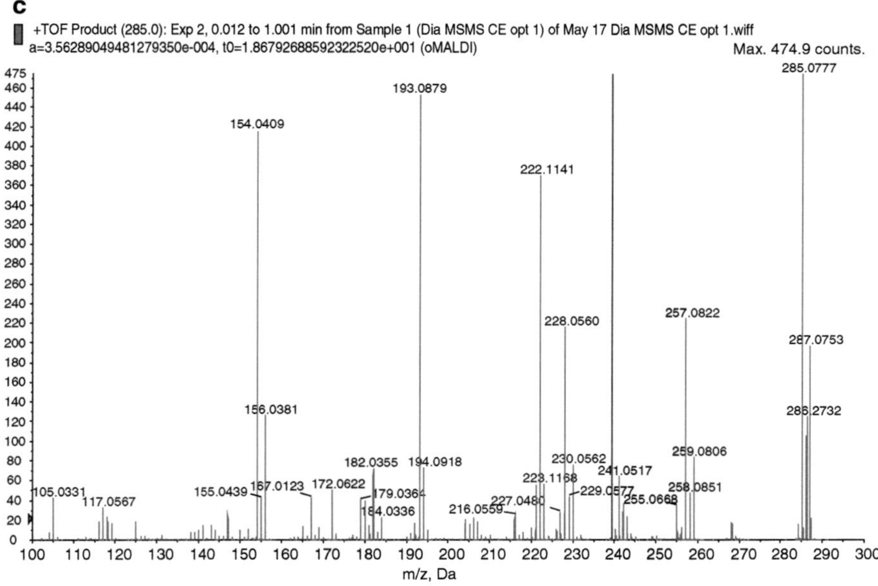

Fig. 14.10 (continued)

14.4.2 *Results*

MS and MS/MS spectra were first acquired from a Diazepam standard using an oMALDI 2 source and a QSTAR Elite System. MS/MS experiments were used to significantly improve both the detection limit and the specificity of detection for a given analyte. The TOF-MS spectrum of Diazepam in Fig. 14.10 illustrates a large number of matrix cluster peaks less than 1,000 Da, indicating the necessity of solving the problem.

A looped MS/MS method was set up with increasing increments of collision energy to identify the major fragment of Diazepam and optimize the intensity for the strongest fragment (Fig. 14.11). Diazepam imaging methods were developed, based on the optimization results for MS/MS methods.

An image of a sagittal rat brain section was then obtained via a low-resolution imaging experiment. This image was obtained at $200 \ \mu m \times 200 \ \mu m$; the typical laser spot size on the oMALDI source is elliptical in shape, measures approximately $100 \ \mu m \times 200 \ \mu m$, and has traditionally been the limiting factor in reducing spatial resolution with the MS imaging method.

To increase the detection limit and specificity, these data were obtained using an MS/MS method optimized for the detection of the m/z 193 fragment of Diazepam (Fig. 14.12).

The resultant "raw data" image was normalized by dividing an image produced from the ion transition from m/z 285 to m/z 193, by an image produced from an ion at m/z 170; the m/z 170 peak was thought to be produced by the fragmentation of a matrix cluster. Both results show that Diazepam has some localization within brain tissues and is not detected in the corpus callosum (Fig. 14.12a).

Following the acquisition of this low-resolution image, a high-resolution MS imaging experiment was set up, using the oversampling method, on a consecutive section of sagittal rat brain tissue. Oversampling involves moving the sample stage in increments that are smaller than the diameter of the laser of the MALDI source during the acquisition of an image. The oversampling technique is illustrated in Fig. 14.13. The pixel size for the high-resolution MS imaging was set at $60 \ \mu m \times 100 \ \mu m$, providing a fourfold increase in spatial resolution. One parameter in the Diazepam MS/MS acquisition method was changed: the accumulation time for the acquisition of individual spectra at each pixel was increased to 500 ms. This increase in spectral accumulation time was used to ensure that the matrix in each pixel was completely ablated or "burned out" (Fig. 14.13).

Figure 14.14 shows the results from high-resolution IMS. The images show a greater degree of detail related to the pathological feature of the tissue section when compared to the optical image (picture) of the same tissue.

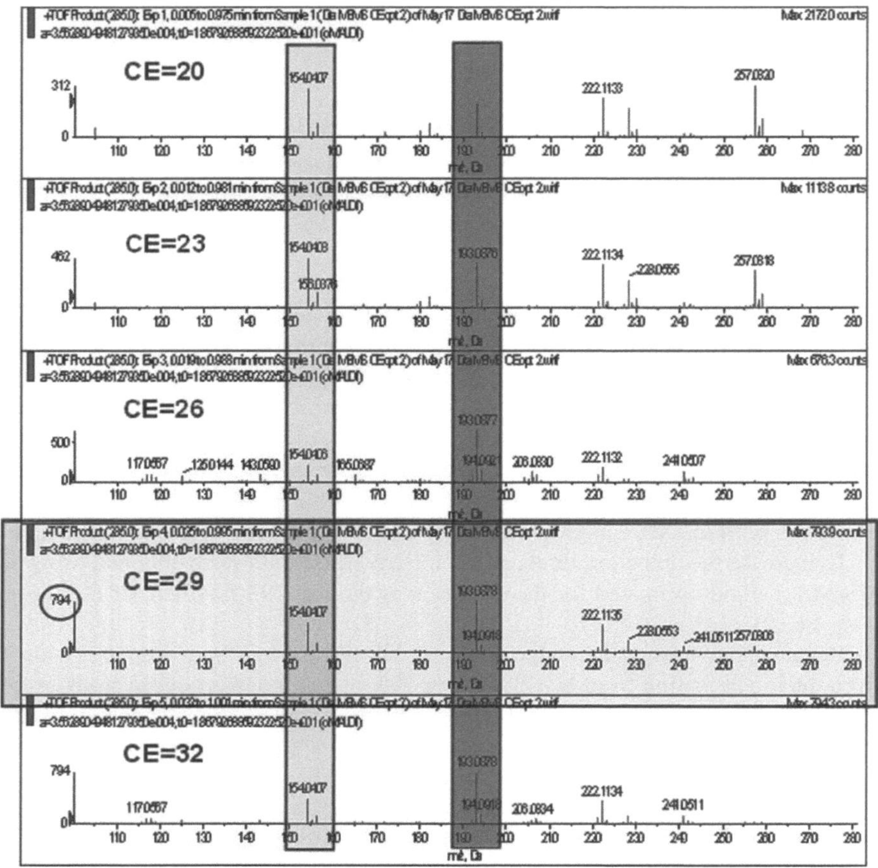

Fig. 14.11 MS/MS optimization experiment for Diazepam. A looped method was designed with five separate MS/MS acquisition methods. Collision energy was varied in 3-V increments from 20 to 32. The 193 fragment gave the highest intensity, at CE=29

14.4.3 Summary

High-resolution IMS was accomplished using the oversampling technique, the execution of which required no changes to the laser optics. The detection of Diazepam in the high-resolution image was consistent with the findings in the low-resolution image. High-resolution IMS experiments provided greater detail vis-à-vis drug distribution in tissue sections, and specific physiological features in the rat brain were identified based on the visualization of Diazepam distribution.

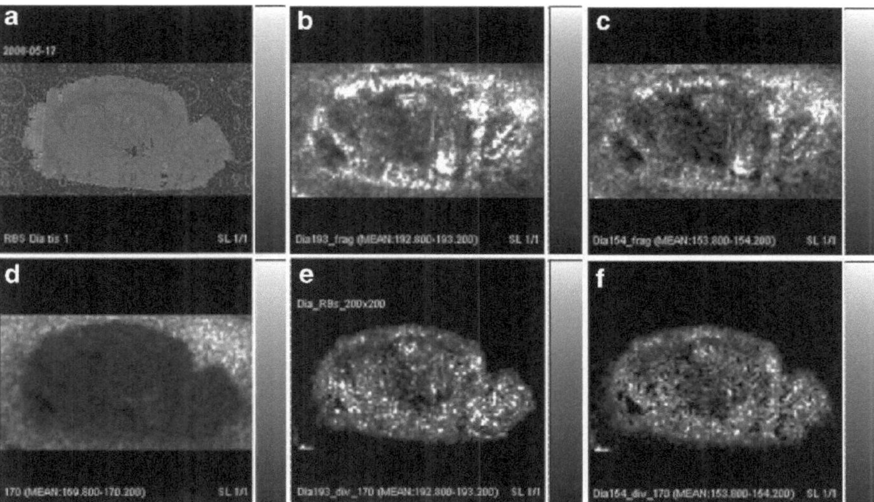

Fig. 14.12 Images of Diazepam at 200 μm × 200 μm resolution. (**a**) Optical image of sagittal rat brain section. (**b**) Raw data MS image of 193 fragment from Diazepam [285 → 193 transition]. (**c**) Raw data MS image of 154 fragment from Diazepam [285 → 154 transition]. (**d**) MS image of $m/z+ = 170$, believed to be a matrix fragment. (**e**) Normalized MS image of 193 fragment from Diazepam [285 → 193 transition]. (**f**) Normalized image of 154 fragment from Diazepam [285 → 154 transition]

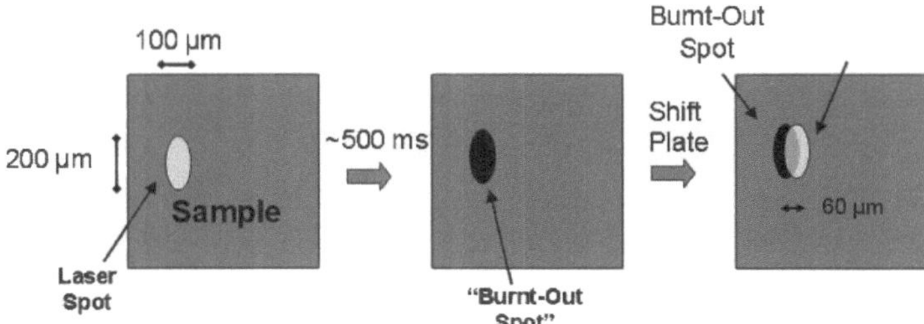

Fig. 14.13 Oversampling schematic. In the oversampling method, the raster increment of the sample stage movement is smaller than the diameter of the laser beam, with complete MALDI ablation at every pixel. The laser spot is elliptical and is 100 μm wide × 200 μm high. Because the spectral signal is dependent upon the matrix, only the fresh region of sample produces an ion current with each overlapping area of ablation; this effectively increases the resolution for an IMS experiment, without altering the laser spot size

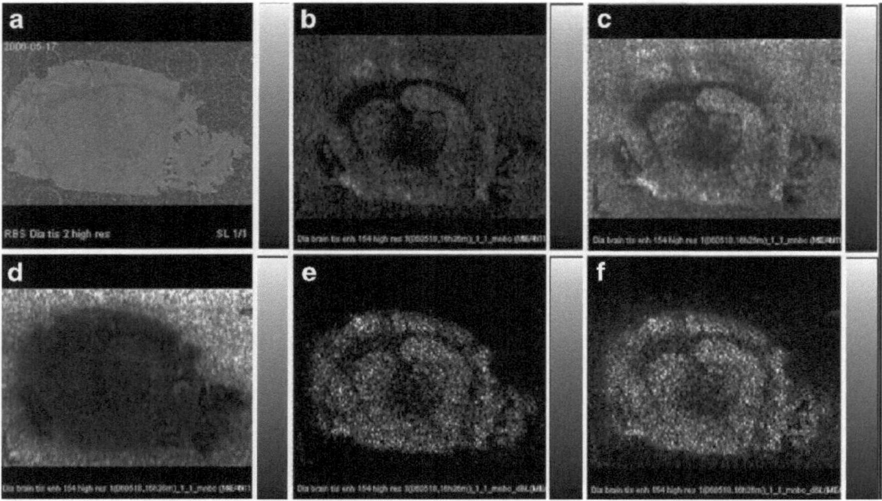

Fig. 14.14 Diazepam images at 60 μm × 100 μm resolution. (**a**) Optical image of sagittal rat brain section. (**b**) Raw data MS image of 193 fragment from Diazepam [285 → 193 transition]. (**c**) Raw data MS image of 154 fragment from Diazepam [285 → 154 transition]. (**d**) MS image of *m/z*+ = 170, believed to be a matrix fragment. (**e**) Normalized MS image of 193 fragment from Diazepam [285 → 193 transition]. (**f**) Normalized image of 154 fragment from Diazepam [285 → 154 transition]

References

1. Stoeckli M, Chaurand P, Hallahan DE, Caprioli RM (2001) Imaging mass spectrometry: a new technology for the analysis of protein expression in mammalian tissues. Nat Med 7(4):493–496
2. Todd PJ, Schhaaff TG, Chaurand P, Caprioli RM (2001) Organic ion imaging of biological tissue with secondary ion mass spectrometry and matrix-assisted laser desorption/ionization. J Mass Spectrom 36(4):355–369
3. Chaurand P, Caprioli RM (2002) Direct profiling and imaging of peptides and proteins from mammalian cells and tissue sections by mass spectrometry. Electrophoresis 23(18):3125–3135
4. Chaurand P, Schwartz SA, Caprioli RM (2002) Imaging mass spectrometry: a new tool to investigate the spatial organization of peptides and proteins in mammalian tissue sections. Curr Opin Chem Biol 6(5):676–681
5. Jurchen JC, Rubakhin SS, Sweedler JV (2005) MALDI-MS imaging of features smaller than the size of the laser beam. J Am Soc Mass Spectrom 16(10):1654–1659

Chapter 15
Bruker Daltonics

Soeren-Oliver Deininger

Abstract In this chapter, a brief overview of running MALDI imaging experiments on Bruker flex instruments is given. General guidelines on the instrument setup and a discussion of the laser diameter settings in general are included. The basic operation of the flexImaging software is described, including the acquisition of the primary optical image, the teaching of the sample carrier, the selection of measurement areas, and the acquisition. The different ways to evaluate the results are also covered: The selection of peaks on a sum spectrum and the intensity distribution thereof, the comparison of selected regions on the tissue to find specific masses, and statistical evaluation, including principal component analysis, hierarchical clustering, and supervised classification, are also covered.

15.1 Preparing Instrument Parameters for MALDI Imaging Experiments

Before the acquisition of the matrix-assisted laser desorption/ionization (MALDI) image, the instrument parameters need to be appropriately set. Most instrument parameters need to be set only once for a given type of experiment; they include mass range and reflectron or linear mode, among others. Here, the sampling rate deserves more attention than in other applications, as it defines how many data points are recorded per second. The sampling rate is directly proportional to the size of the spectrum. A spectrum recorded with a sampling rate of 1 S/s will be 10 times larger than a spectrum with 0.1 GS/s. The chosen sampling rate should be as small as possible without compromising the peak shape in the spectrum; this is because for an imaging dataset with many thousands of spectra, the size of the spectra accumulates quickly. For a linear measurement of proteins, a sampling rate of 0.1 GS/s is usually sufficient.

S.-O. Deininger (⊠)
Bruker Daltonik GmbH, Fahrenheitstrasse 428359, Bremen, Germany
e-mail: Soeren-Oliver.Deininger@bdal.de

M. Setou (ed.), *Imaging Mass Spectrometry: Protocols for Mass Microscopy*,
DOI 10.1007/978-4-431-09425-8_15, © Springer 2010

15.1.1   Laser Diameter Settings

The smartbeam laser of flex instruments has an adjustable focus size that can be adjusted in several steps (e.g., ultralarge, large, medium, small, and minimum). The question of what exactly the diameter of the laser focus should be for each setting is often asked; unfortunately, that question is not easily answered. The diameter of a Gaussian-shaped laser beam profile is typically reported as full-width at half-maximum. Although the smartbeam laser has no Gaussian beam profile, the energy profile averaged over many shots has a Gaussian envelope. What matters in terms of the imaging experiment is the "effective" laser diameter that affects the sample; this effective diameter is strongly dependent on the laser energy. If, for example, a high laser energy level is needed, the effective diameter will be larger than with a lower laser energy level. The effective diameter is therefore also difficult to estimate; it can really only be estimated by acquiring spectra at a given position, moving the laser a defined distance, measuring again, and then looking at the distance at which the spectra still have comparable intensities. The recently introduced smartbeam II laser has a flat-top envelope and is therefore more robust in MALDI imaging.

In terms of MALDI imaging, the salient question is which lateral resolution can be achieved under practical conditions with real tissue. Here, we have observed that resolutions down to the 70-μm level can be obtained with the "large" laser setting and a sinapinic acid matrix, which is best suited for proteins. With a "medium" laser setting, ~50-μm resolution is possible with sinapinic acid. With a cyano-4-hydroxy cinnamic acid matrix, the "small" laser setting allows for an imaging resolution of approximately 25 μm. These values should be considered only guidelines, because they depend upon the laser energy, and the necessary laser energy in turn depends upon the matrix, the preparation, and the amount of salt and lipids in the tissue.

15.2   Data Acquisition and Processing with the Bruker Daltonics MALDI Imaging Solution

The core of Bruker Daltonics's software for MALDI imaging is the flexImaging program. It is used both as an editor to set up the imaging run and to visualize and evaluate results.

flexImaging requires an optical image of the sample, which is referred to as the "primary image."

15.2.1   The Primary Image

This optical image will later be used to define the measurement areas, and the results will be overlaid with this primary image. In setting up a new imaging sequence, the user is asked at one point to register this image in the instrument coordinates; this is done by aiming with the crosshair of the instrument camera onto a specific feature and defining the corresponding feature in the optical image. Whether this primary image needs to be taken before or after the application of the matrix depends on how the matrix was prepared. If the matrix was prepared as a homogeneous layer – for example, by the ImagePrep station or a spray preparation – the primary image should be taken before matrix preparation. In this case, it is also necessary to put some teach marks onto the sample that can be clearly recognized under the matrix layer. Liquid Paper ("white-out") has been shown to work nicely here.

If the sample has been prepared by spotting a grid of matrix droplets through the use of a robot, then the primary image needs to be taken with the matrix spots already on the sample, because the sample will be "teached" onto the corner positions of the grid.

For the acquisition of the optical image, a photographic slide scanner with an adaptor for microscopy slides is most convenient. Office scanners could also be used; however, the quality of the optical image is not as good as from a slide scanner. Microscopes can also be used, but the field of view is often too small to obtain an image of the entire sample.

15.2.2 Creating a New Imaging Sequence

If a new imaging experiment is being set up, first it is necessary to introduce the sample into the instrument. In flexImaging, the option to set up a new sequence is selected. The initial setup of the sequence is now controlled by the new sequence wizard; this wizard guides the user interactively through the necessary steps. First, a name for the sequence must be defined. On the next page, the wizard asks for the imaging mode; the user can choose from three modes of operation that are defined by the way the matrix is applied to the sample. In the case of a homogeneous matrix layer, the user must also select the option "uniformly distributed coating." In this workflow, the user can arbitrarily choose the lateral resolution of the image in the measurement region. In the case of a robotically spotted matrix grid, the user chooses the "spot microarray" option; in this workflow, the lateral resolution of the imaging experiment is defined by the center-to-center distance between matrix spots, and so the user must "tell" the software at this point how many rows and columns are present or what are the center-to-center distances in the x- and y-directions. Alternatively, it is also possible to name a file with spot coordinates from the robot. The third workflow, "tissue profiling," is used only if specific spots at arbitrary positions are being measured.

Next, the wizard asks for the primary optical image. After this is selected, the process of teaching begins. In this teaching, the optical image is correlated with the sample on the sample carrier. For three reference spots, the corresponding features are first selected with the crosshair of the MALDI camera optic and on the primary optical image. After this teaching is finished, the new sequence wizard closes and the primary optical image is displayed in flexImaging. Because of the teaching, the software can now correlate each pixel on the primary image with the respective position of the sample carrier. flexImaging can now be used to remote control the sample carrier, to move to a specific point and take a spectrum, that is, check the quality of the preparation or estimate the appropriate laser energy level for the measurement (Fig. 15.1).

As a next step, the measurement region must be defined, either as a rectangular or as a polygon-shaped region. It is possible to define more than one region on the same sample. In the region's pane in flexImaging, the lateral resolution for the region can be defined, as well as the parameters for autoXecute, which defines the automatic measurement and processing. Setting up for automatic measurements is simple, because the laser is operated at a fixed laser energy level with a fixed number of shots to be accumulated. Once the measurement regions are selected, the measurement can start. The measurement can be triggered either for one sample directly from flexImaging or, if several samples are located on the same slide, by a batch-acquisition queue.

There are special considerations for mass spectrometry (MS)/MS imaging. Targeted analysis – such as small-molecule drug imaging – can be performed on time-of-flight (TOF)/TOF instruments; in this case, the only difference between TOF/TOF and an MS-only image is the fact that a parent mass must be selected as sequence property and an MS/MS autoXecute method must be chosen for the acquisition. It is recommended that this acquisition be undertaken in "fragments only" mode. The evaluation will be carried out later, on a characteristic fragment ion.

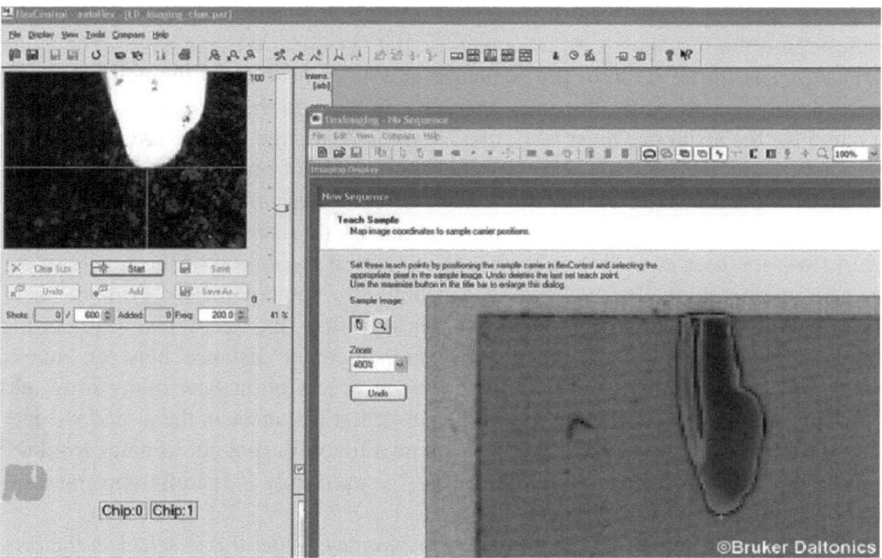

Fig. 15.1 Teaching during the setup of a run. The camera picture of the flex instrument is seen on the *left side* and the new sequence wizard of flexImaging with the optical primary image of the sample on the *right side*. The teaching is done by positioning the crosshair of the camera onto a specific point and then selecting the same point on the optical image. In this case, the reference point has been painted with Liquid Paper so that it is visible under the matrix layer

15.3   Evaluation of the Results

15.3.1   *Distribution of Mass Signals*

Once the measurement is finished, flexImaging automatically loads the imaging data. Also, during the acquisition, it is possible to load the already acquired data into flexImaging. flexImaging offers the option to read the raw spectra, the processed spectra, or peak lists. To load the processed spectra, it is recommended that baseline subtraction and smoothing be performed during acquisition.

By default, flexImaging displays an overall average spectrum for the dataset in the spectrum display. A mass marker can be used to select peaks in the overall spectra; at the same time, the intensity distribution of this signal will be shown in the image display as a color intensity (Fig. 15.2).

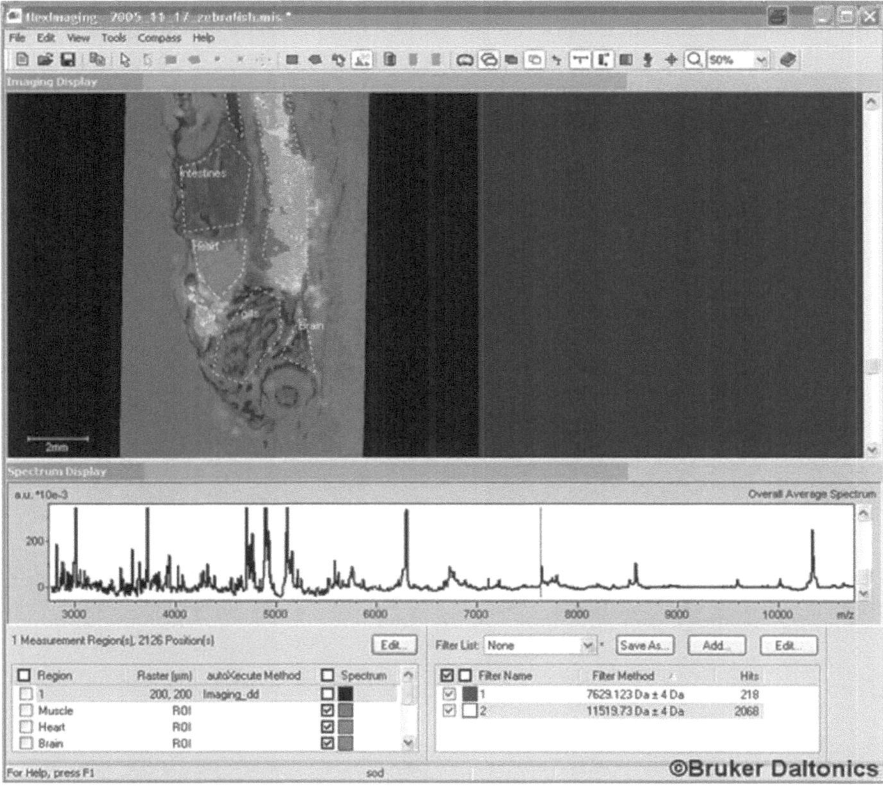

Fig. 15.2 Overview of flexImaging. In this case, an intensity map of the selected mass is displayed as an overlay over a co-registered, histologically stained whole-body section of a zebrafish. Different regions of interest have been defined on the zebrafish. The masses can be selected by clicking on the respective signal in the overall average spectrum as well as average spectra from different regions

The following information can be shown in the image display. The primary image can be switched on or off, a co-registered secondary image can be switched on or off, and the measurement region can be displayed; it is also possible to display regions of interest, the teach position, a scale bar, and the raster positions. In some cases, the primary image will show sufficient contrast to make the recognition of specific regions possible. For example, with brain sections, white or gray matter is easily recognized on an unstained section. In many cases, however, a detailed analysis of histology will require histological staining. For this reason, it is possible to co-register an image. This image could be a histological stain of a consecutive section, but it is also possible to remove the matrix layer after measurement with 70% ethanol and perform histological staining. In either case, this image can be co-registered in the "co-register image" dialog box (Fig. 15.3).

Here, the primary image is shown on the left side, with the image to be co-registered on the right side. The registration is done by selecting three corresponding spots on each image. It is now possible to overlay the histological image with the MS image. To get a clearer idea of the localization of the results, it is also possible to cross-fade between the two images. Usually, an image is not just analyzed by looking at the distribution of every single peak, because one wants to see characteristic differences between defined regions of the sample, for example, which peaks are present only in the tumor but not in the normal epithelium.

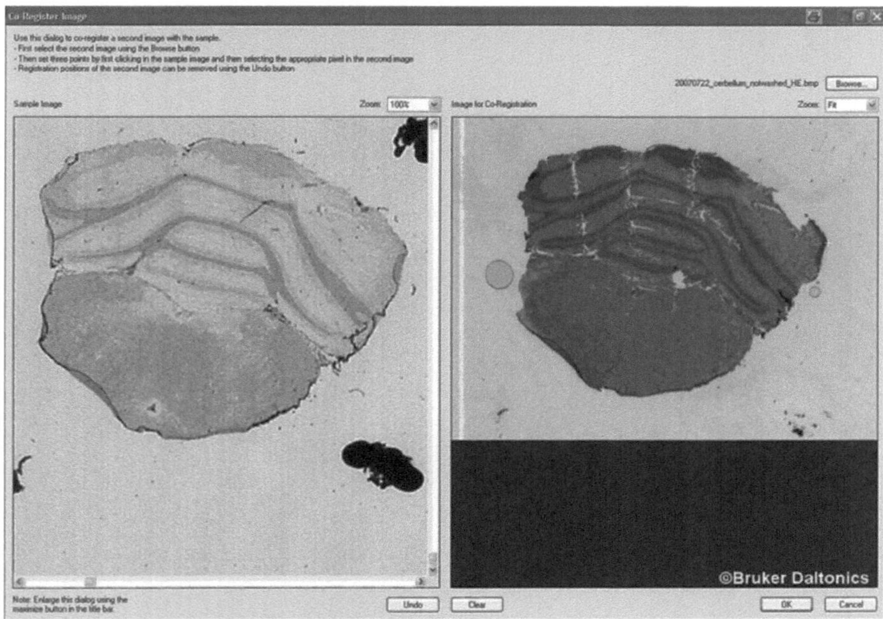

Fig. 15.3 The "co-register image dialogue." The primary image is on the *left side*; on the *right side* is the histologically stained image that will be co-registered. The co-registration is done by selecting three corresponding spots on each image

In this case, one defines regions of interest. These regions can also be named, accordingly, "tumor" and "epithelium." In the region pane in flexImaging, the average spectra for each region can be switched on via a checkbox. The average spectra of the selected regions can now be compared in the spectrum display, either as a pseudo-gel-like display or as an overlay of mass spectra traces. In this view, the differences between the regions become apparent. The peaks that are different can now be selected for display in the imaging display to confirm the localization.

Interesting signals can be added to a result filter list, for later reference. Each mass filter can be selected or deselected, so different masses can be overlaid. If two mass filters overlie, then a mixed color is created. For each mass filter, adjustments, such as the upper and lower intensity limits, can be selected.

15.4   Statistical Analysis

flexImaging interacts with ClinProTools for the statistical evaluation of data. The statistical analysis has two aspects: the analysis of one dataset and the comparison of different datasets.

15.4.1   Statistical Analysis of Single Imaging Datasets

15.4.1.1   Principal Component Analysis

Principal component analysis (PCA) is one way to reduce multidimensional data to lower dimensions while retaining as much information on the dataset as a whole as possible. In the context of MALDI imaging, this leads to a situation where similar spectra are displayed with similar color intensity. This is the preferable way of performing an unsupervised feature extraction, so a user of MS can generate images that reflect the overall variance of the dataset without having a complete understanding of the underlying histology. It can usually be assumed that the first principal components (PCs) will reflect the histology of the section, if a homogeneous sample preparation has been done.

Each mass spectrum with n-peaks can be considered a data point in an n-dimensional coordinate system. The PCA now transforms this coordinate system in such a way that the axis of the new coordinate system points in the direction of the largest variance in the dataset. The axes of the new coordinate system are called PCs. The first PC points in the direction of the largest variance in the dataset, the second PC points in the direction of the second-largest variance, and so on. The numerical value of a data point on a PC is called a score, and the score for each mass spectrometric data point can therefore be translated back to a color intensity on the MALDI image.

To perform PCA practically with ClinProTools, the raw spectra of the entire dataset must be loaded into ClinProTools. Once they are displayed, the calculation can be immediately initiated by clicking the "PCA" button. At this point, ClinProTools optionally aligns the spectra and performs the calculation. Once the calculation is finished, the result is displayed as a window with scores and loadings plots. More importantly, at the same time, a file is written that contains the results of the analysis; these results can now be imported into flexImaging by using the "add" button in the result filter list. A wizard then guides the user through the process. If individual PCs are selected, the rainbow mode is recommended for the display, because this will exaggerate the contrast on the image.

As a second option, it is also possible to import not the PC scores but rather the masses that contribute most to the variance in a dataset.

15.4.1.2 Hierarchical Clustering

Other unsupervised classification, such as hierarchical clustering, enhances the IMS date interpretation. Here, the underlying principle is that the more similar two spectra are, the closer they are in the multidimensional data space; classification is based on this distance. A hierarchical tree always contains all data points, and each branch in such a tree can be considered a class. The important question here is which classes are relevant; with flexImaging, this can be estimated in a interactive way. After the data were loaded into ClinProTools, a clustering calculation can be started by clicking on the "hierarchical cluster" button. Here, the option "create full tree" should be used. In the advanced options, it is possible to select different distance metrics. After the calculation is finished, the cluster is displayed; at the same time, the result of the clustering is written into a file. This file can now be imported into flexImaging using the "add" button in the result pane. A wizard guides the user through the process. In flexImaging, the top nodes of the tree are displayed in a tree browser. By selecting different branches on this tree, the respective data points on the image are highlighted. The tree is browsable, so it is possible to browse to the branch that contains the spectra specific for histological structures (Fig. 15.4).

15.4.1.3 Statistical Analysis on Many Datasets

Although the statistical analysis of single datasets is very useful in discovering the various features of a dataset, it will usually be necessary to compare different datasets. A common task would be to find markers that separate them – for example, one type of tumor from another type of tumor. In this case, it is necessary to generate MALDI images from a larger set of different patients first. Next, the regions that contain the tumor spectra need to be defined as regions of interest. The spectra for each region of interest can now be exported from a spectra list. It is necessary, for example, to group the spectra of tumor type one into one class and tumor type two in the second class; please refer to the ClinProTools manual on how to perform this.

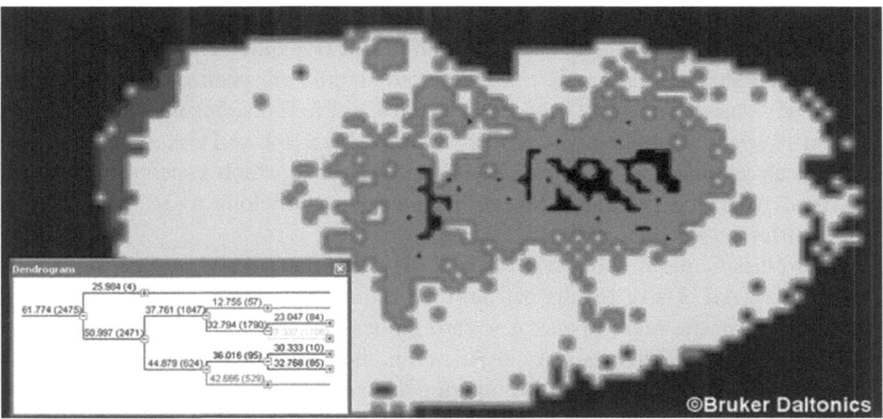

Fig. 15.4 Display of hierarchical clustering result in flexImaging. Branches of the hierarchical tree can be selected in different colors, and the respective pixels in the image will be shown. In this example, the clustering is done on a rat kidney dataset

In ClinProTools, the two classes are then loaded. It is necessary to use the "support spectra grouping" switch in the data preparation settings of ClinProTools; each patient will then be considered one sample while each spectrum will then be considered a replicate of that sample.

ClinProTools does offer considerable information on the two classes. The average spectra for each class will be calculated and, for each peak, it is possible to display the average and standard deviation for each class. It is also possible to display a box-and-whiskers plot or a receiver operator characteristic for each peak. The spectra can also be visually compared in the pseudo-gel view. At this point, it is easy to look for those peaks that separate the two classes. Note that this "supervised" approach is very different from a PCA; with the supervised approach, the focus is set entirely on those peaks that separate the classes, whereas in a PCA, the focus is on variance in the dataset. Especially when samples from different patients are being compared, it can be expected that the overall variance in the dataset is a patient-to-patient variance; a PCA, therefore, would not separate the two classes but rather each patient from one another.

After the masses of interest are located as candidates, it becomes easy to validate findings on the imaging datasets by comparing that they are indeed specific, for example, to a cancer type in question.

15.4.2 Supervised Classification

As described in the previous section, a dataset can be used to generate a software model that can later be used to classify unknown spectra from different imaging runs.

Different algorithms can be used here: a support vector machine (SVM), a supervised neuronal network (SNN), a genetic algorithm (GA), or a quick classifier (QC).

The generated models can be used to classify unknown spectra. Here, it is useful to use the "classify in batch mode" option in ClinProTools, because an unlimited number of spectra can then be classified. The SVM, SNN, and GA algorithms will generate an output that will simply advise of the class to which a spectrum has been classified. With the QC algorithm, each spectrum can belong to all classes, each with a different weight.

After classification, ClinProTools asks the user where the classification results should be saved. These results can now be imported into flexImaging using the "add" button in the region pane; a wizard guides the user through the process. In the case of a SVM, SNN, or GA classification, the resultant image will display the classification, and each pixel in the imaging dataset will show the same color intensity. In the case of a QC classification, the weight value will be translated into a color intensity.

Chapter 16
Imaging and Molecular Identification of Biomolecules on Tissue Sections with AXIMA-QIT: Shimadzu Corporation

Masaru Furuta, Yuki Sugiura, and Mitsutoshi Setou

Abstract In MALDI-IMS experiments, because of the complex MALDI process on the tissue surface, the ability to both visualize and identify molecules directly on the tissue surface by tandem mass spectrometry (MS^n) is essential. Here we introduce an imaging system based on a MALDI-quadrupole ion trap time-of-flight type instrument (AXIMA-QIT; Shimadzu Corporation), which is compatible with both imaging and highly sensitive MS^n. In this chapter, we present the attractive features of the AXIMA-QIT as an imaging instrument and introduce the variety of application topics. The visualized biomolecules were successfully identified by MS^n directly on the tissue surface, with a strong ability to isolate precursor ions. The presented IMS system with AXIMA-QIT covers a wide range of molecular imaging including phospholipids and other endogenous metabolites and also tryptic-digested proteins in various biological samples.

16.1 Introduction

Until today, imaging mass spectrometry (IMS) studies have been conducted on a variety of topics, including biological applications [1, 2] and pathological applications [3, 4]. In addition to the analysis of protein described in the foregoing references, analysis of small molecule in mammalian tissues has also been conducted [5, 6], and histopathological materials [7, 8] and pharmacokinetics in rat whole-body sections [9] have been studied.

M. Furuta (✉)
Applications Development Center, Analytical Applications Department, Analytical and Measuring Instruments Division, Shimadzu Corporation, 1 Nishinokyo-Kuwabaracho, Nakagyo-ku, Kyoto 604-8511, Japan
e-mail: furu@shimadzu.co.jp

Y. Sugiura and M. Setou
Department of Bioscience and Biotechnology, Tokyo Institute of Technology, 4259 Nagatsuta-cho, Midori-ku, Yokohama, Kanagawa 226-8501, Japan

Y. Sugiura
Department of Molecular Anatomy, Hamamatsu University School of Medicine, 1-20-1 Handayama, Higashi-ku, Hamamatsu, Shizuoka 431-3192, Japan

M. Setou (ed.), *Imaging Mass Spectrometry: Protocols for Mass Microscopy*,
DOI 10.1007/978-4-431-09425-8_16, © Springer 2010

In general, these imaging instruments were based mostly on matrix-assisted laser desorption/ionization (MALDI)-time-of-flight (TOF), MALDI-TOF/TOF, or TOF-secondary ion mass spectrometry (SIMS). In particular, MALDI-TOF imaging instruments are high throughput, and TOF-SIMS imaging instruments can provide submicrometer spatial resolution (>500 nm). Both instruments have unique and powerful advantages in IMS. However, there is a potential problem with respect to the identification or structural analysis of molecules of interest. When the goal is to identify visualized molecules, it is essential to use a tandem mass spectrometer that can efficiently separate precursor ions in tandem mass spectrometry (MSn). With all these issues in mind, we developed an imaging system based on a MALDI-quadrupole ion trap (QIT)-TOF (AXIMA-QIT; Shimadzu, Kyoto, Japan) [10] mass spectrometer (Fig. 16.1). The QIT acts as both an ion separator and a storage machine. Therefore, the instrument can provide molecular identification with

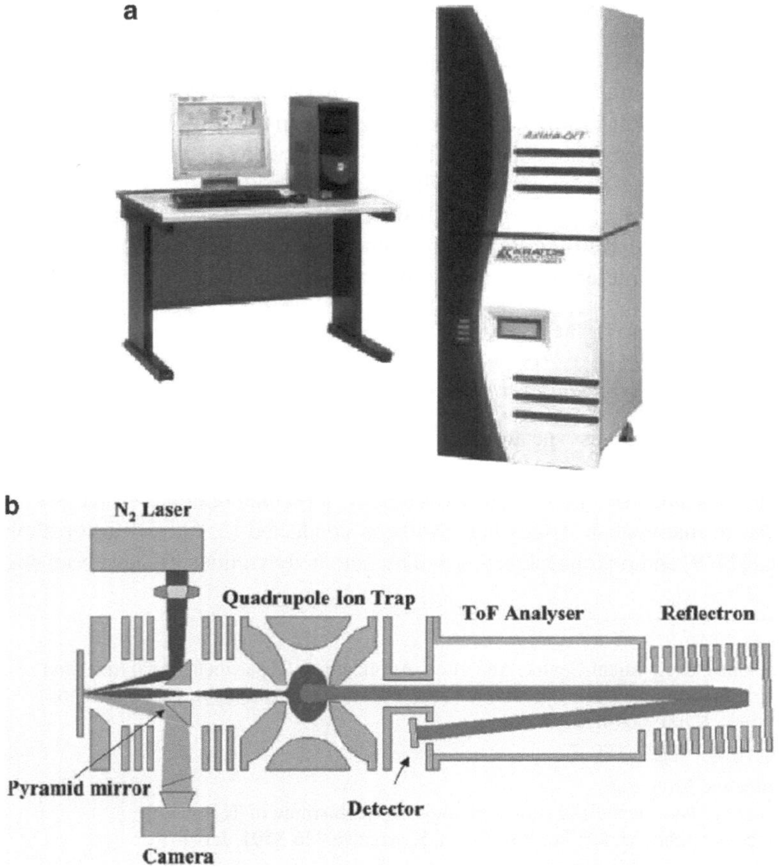

Fig. 16.1 Overview of imaging mass spectrometry (IMS) system with AXIMA-QIT (**a**) and instrumental configuration of MALDI-QIT TOF-MS (**b**) (From [10], in part) (Reprinted from Martin et al., Rapid Commun Mass Spectrom 17:1358–1365.)

highly sensitive MS^n from a complex mixture of ions generated on the tissue surface. In this chapter, we describe the IMS demonstrations under which AXIMA-QIT can be used for molecular imaging.

16.2 Imaging and Identification of Lipids in Negative Ion Detection Mode

The AXIMA-QIT IMS system provided good capabilities in both molecular imaging and identification. As a first example, we performed visualization of negatively charged lipids, e.g., phosphatidylinositol (PI) and sulfatide (ST), in a mouse brain section with use of 2,5-dihydroxybenzoic acid (DHB) matrix. Figure 16.2 shows a photograph of the imaging region in the mouse cerebellum section (Fig. 16.2a) and the representative mass spectrum obtained (Fig. 16.2b). Here, we selected two relatively minor mass peaks for detail analysis, which were at m/z 885.5 and 904.7. Images of these two m/z values were reconstructed (Fig. 16.2c), and we found that these peaks were distributed complementarily i.e., m/z 904.7 was in the white matter and medulla oblongata, and m/z 885.5 was in the gray matter. To identify these peaks, MS^n was performed (Fig. 16.2d).

In the MS^n analysis, from the high-quality product ion spectra obtained with QIT-TOF, characteristic peaks to determine molecular species were detected. The peak at m/z 241 was one deprotonated cyclic inositol monophosphate, and the peaks at m/z 283 and 303 were deprotonated fatty acids C18:0 and C20:4, respectively (Fig. 16.2d, upper panel). The obtained product ion mass spectrum indicated that m/z 885.5 was derived from PI, and the composition of fatty acid was 1-stearoyl-2-arachidonoyl [PI (18:0–20:4)]. In another product mass spectrum of Fig. 16.2d (lower panel), the peak at m/z 540 was derived from lyso-ST. This lyso-ST was detected when ceramide contained hydroxy fatty acid. Therefore, the fatty acid of ST could be calculated from the difference in m/z between the precursor ion and lyso-ST. We concluded that m/z 904.7 was ST (d 18:1–C 24 h:1).

As shown, imaging with an AXIMA-QIT tandem mass spectrometer allows visualization of not only the differences in phospholipid classes but also the differences in fatty acid composition (i.e. molecular species).

16.3 Imaging of Positively Charged Lipids Within the Minute Structure of a Mouse Retina

Next, we demonstrate that the IMS system is capable for lipid imaging even within minute structures of biological samples.

The retina in the eye has an extremely thin layered structure, a nine-layered structure within a width of about $150\,\mu m$ [11]. In the following demonstration, we visualized the layer-specific distribution of phospholipid molecular species.

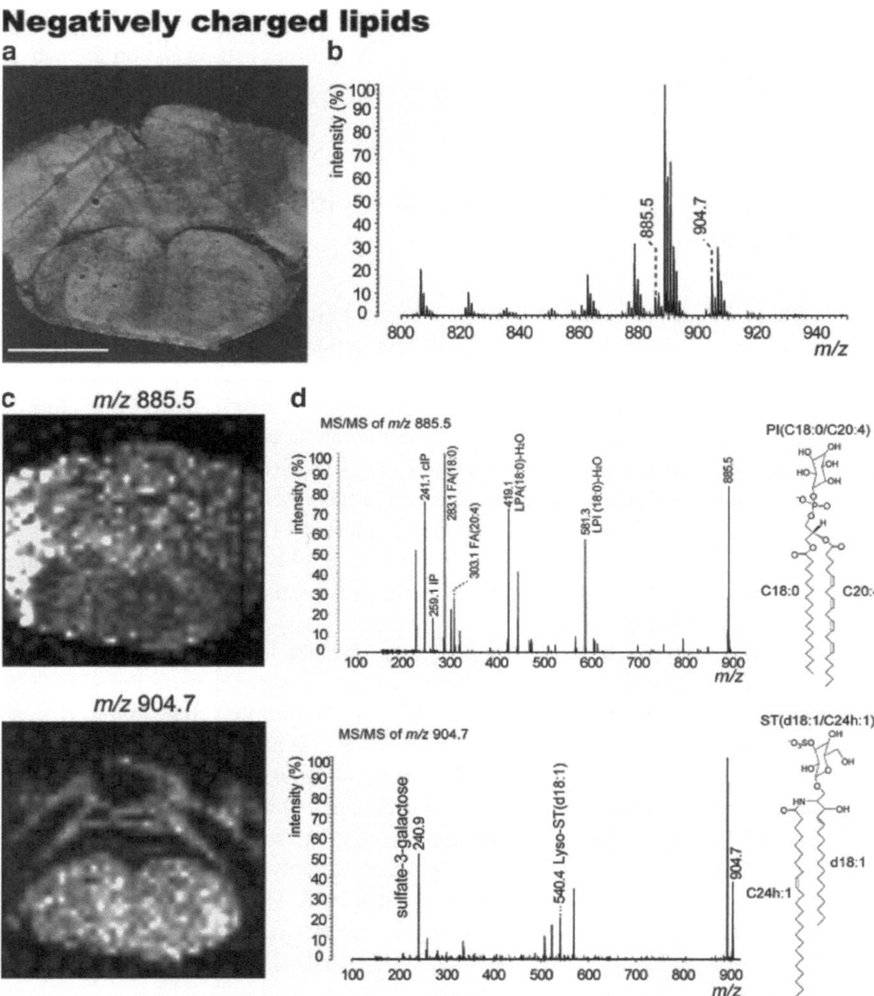

Fig. 16.2 Example of lipid imaging and precursor ion mass spectra with negative ion detection mode. From the imaging region (**a**), negatively charged lipids based on the head group structure were detected (**b**). The molecules of *m/z* 885.5 and 904.7 were distributed in the gray matter and white matter, respectively (**c**). Precursor ion mass spectrometry showed that *m/z* 885.5 was phosphatidylinositol (18:0–20:4) and *m/z* 904.7 was sulfatide (d 18:1–C 24 h:1). *cIP* cyclic inositol phosphate, *IP* inositol phosphate, *FA* fatty acid, *LPA* lysophosphatidic acid, *LPI* lysophosphatidylinositol (**d**). *Bars* 2.0 mm (Reprinted from Shimma et al., Anal Chem 80(3):878–885.)

The retina sections were sliced at a thickness of 8 μm from a mouse eyeball adapted to the dark and sprayed with a DHB matrix solution; IMS was then performed at a scan pitch of 50 μm. In the spectrum obtained, several peaks with higher intensities were observed, and we identified that most of them were derived from phosphatidylcholine molecular species [12] (Fig. 16.3, top panel). The reconstructed ion

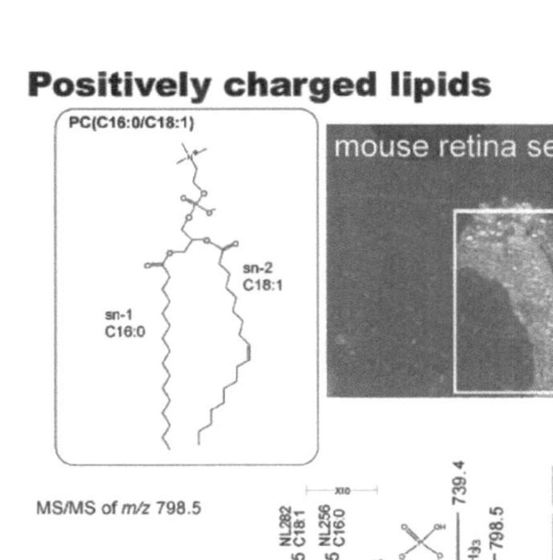

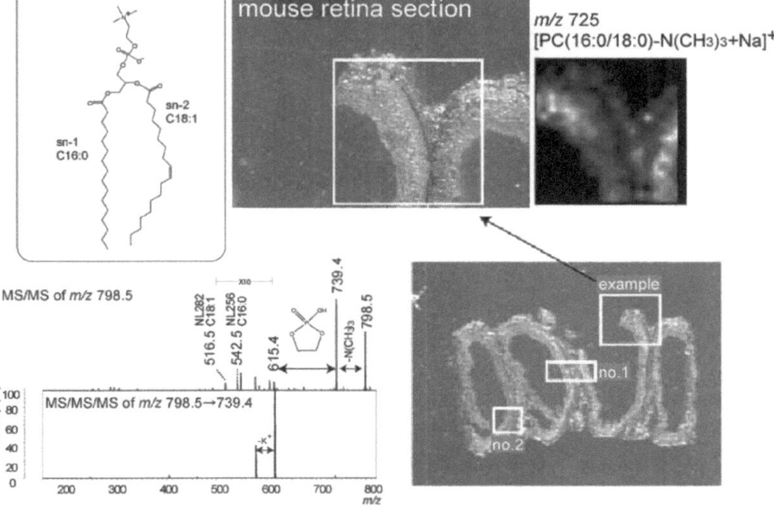

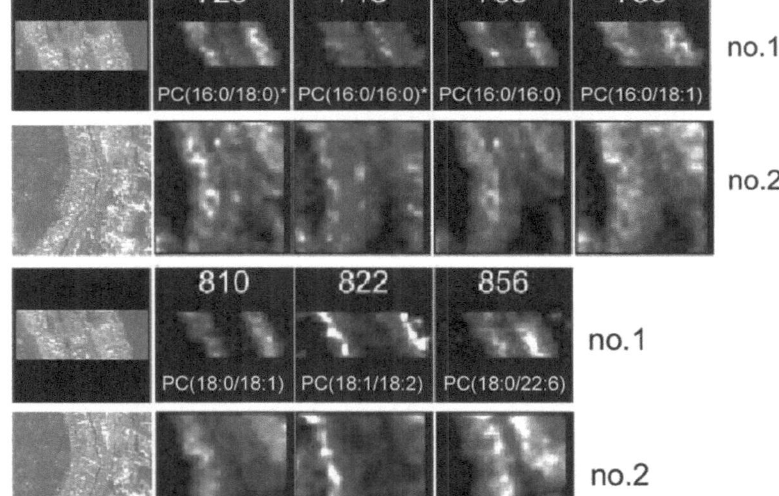

Fig. 16.3 Example of lipid imaging in the minute layer structure of the mouse retina section. *Top panel:* An abundant class of phospholipids, which is phosphatidylcholine, *(PC)* is visualized and identified on the mouse retina section. *Bottom panel:* The imaging results of various molecular species of PCs; i.e., PC with different fatty acid composition reveals that they are rather heterogeneously distributed among the retinal layer structures. The molecular species of these peaks were identified by direct MS^n. Scan pitch, 50 μm

images showed that although PC (diacyl-16:0/18:1) widely distributed among the optic fiber layer, the ganglion cell layer, and the inner plexiform layer, some molecular species of PC showed characteristic layer-specific distributions; e.g., PC (diacyl-16:0/16:0) distributed in the inner nuclear layer and the outer plexiform layer on the retina section. Moreover, PC (diacyl-18:0/22:6) distributed in the outer and inner segments. These results demonstrate the capability of AXIMA-QIT instrument for use as a mass microscope (i.e., a IMS with higher spatial resolution than that of the naked eye).

The visualization of lipid distribution in situ is very important in molecular and functional biology, especially in relationship to diseases. Although some lipid imaging methods have been conducted by the use of tags [13] and interaction between lipids and other biomolecules [14, 15], these procedures provide visualization only of specific classes of lipids. However, as shown above, visualization of the phospholipids reveals a quite different distribution on the tissue section among their classes and even their molecular species. So far, only IMS can visualize the distinct tissue distribution of phospholipid species that have different lipid moieties (*sn*-1 or *sn*-2), e.g., C16:0 and C18:0. Therefore, visualization and direct identification using IMS and direct tandem MS, especially with a high-performance instrument such as AXIMA-QIT, will soon become an essential tool for studies on lipids and related metabolites.

16.4   Visualization of Heme B localization in Colon Cancer Liver Metastasis

In another example, we performed visualization of distinct endogenous metabolite expression in a human cancerous tissue section [7]. In this case, we performed IMS of the disease liver section with metastatic colon cancer, in positive detection mode. The representative mass spectrum in the imaging region is shown in Fig. 16.4b. From the spectrum, we focused and reconstructed the ion distribution image at m/z 616 (Fig. 16.4b, arrow). Comparing histological images with the ion distribution map, we found that the intensity of m/z 616 in the cancerous region was significantly lower than that in the normal region.

To identify the molecule for the ion at m/z 616, we performed MS^2 and further MS^3 on the liver section. Considering the m/z value and peak pattern in MS^2 and MS^3 product ion spectra, it was found that m/z 616 was generated from heme B; neutral losses observed were assigned to loss of CH_2CH_2COOH (73 Da), CH_2COOH (59 Da), or $COOH$ (45 Da), respectively. Heme B consists of an iron atom and porphyrin and is known as a prosthetic group in hemoglobin, which is a protein in erythrocytes. The presented results indicate the difference of molecular composition between the blood-rich organ liver tissue and the ischemic metastatic colon cancer [16, 17].

From the analytical aspect, this case clearly indicates that MS^3 is essential for the accurate identification of molecular species. As the foregoing example shows, lipids that contain a phosphocholine (e.g., phosphatidylcholine and sphingomyelin)

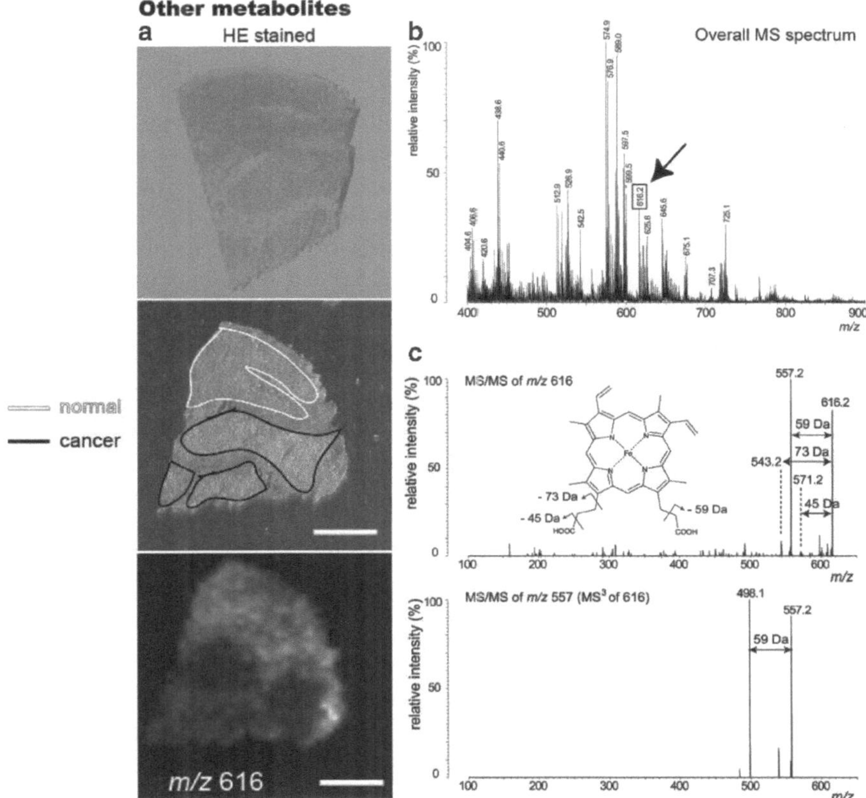

Fig. 16.4 Imaging of distinct localization of heme B in the colon cancer liver metastasis. In the optical image (**a**), cancerous and normal regions are indicated. *Bars* 1 mm. The time required for IMS of a 4×4 mm^2 region was about 112 min (1,681 data points). Overall mass spectrum (**b**) and product ion spectra of MS^2 and MS^3 (**c**) are shown. The *m/z* value and fragment patterns indicate that the product ion of m/z 616 is heme B. Consecutive neutral losses of 73, 59, and 45 Da correspond to CH$_2$CH$_2$COOH, CH$_2$COOH, and COOH, respectively. The molecular structure of heme B is also shown (Reprinted from Shimma et al., J Mass Spectrom Soc Jpn 55:145–148.)

showed a product ion mass spectra with neutral loss of 59 Da (corresponding to trimethylamine) and 183 Da (corresponding to trimethylamine and cyclophosphate). Therefore, performing only MS^2 analysis may increase the risk for incorrect molecular assignment; we could not refute the hypothesis that the peak at *m/z* 616 might be derived from lyso-phosphatidylcholine. In this regard, high performance in MS^n ($n > 2$) analysis provided by the MALDI-QIT-TOF instrument enables us to perform easy and reliable molecular identification, which is unavailable in MALDI-TOF or MALDI-TOF/TOF instruments. In this example, by performing MS^3, we confirmed another neutral loss of 59 Da (see Fig. 16-4c) and thus successfully performed the correct molecular identification.

16.5 Imaging and Identification of Peptides in Mouse Brain Sections

Finally, we present the QIT-TOF instrument's ability to analyze peptide ions directly on the tissue sections. We utilized trypsin-digested mouse brain sections [18, 19] to evaluate the instrument's performance for both visualization and molecular identification of peptides. We performed IMS and MS^n in positive ion mode. Figure 16.5 shows the results of the IMS experiment, including a photograph of a mouse cerebellum section (Fig. 16.5a) and representative mass spectrum (Fig. 16.5b). Images of two m/z values, 1,460.8 and 1,743.9, were reconstructed (Fig. 16.5c, d), and we found these peaks distributed complementarily: m/z 1,460.8 was in the white matter and medulla oblongata, and m/z 1,743.9 was in the gray matter. To identify these peaks, MS^n was performed directly on the tissue sections, and they were identified as fragment ions of myelin basic protein (MBP) and histone H2B, respectively (Fig. 16.5e, f).

16.6 High-Performance MS^n on the Tissue Surface with AXIMA-QIT

To perform proper molecular identification with MS^n on tissue sections that have extremely complex mixtures of biomolecules, precursor ion selection with high resolution is a very important issue. Moreover, a highly efficient fragmentation process is required to obtain significant amount of product ions especially under performing precursor ion selection at high mass resolution. In this regard, two instrumental features of AXIMA-QIT enables successful molecular identification on the tissue sections: the ions generated on the tissue surface do not enter the TOF analyzer directly, and before entering the TOF analyzer, ions are stored and energy is cooled in the QIT. Therefore, QIT can cancel the difference in ionization position and initial energy. Furthermore, even if high-resolution precursor isolation was performed, the S/N and mass resolution tended to be unchanged by the QIT equipment.

Figure 16.6 demonstrates that AXIMA-QIT enables us to isolate a single monoisotopic peak with higher resolution (1 Da) than that of a TOF/TOF instrument. Employing an extra-high separation mode, the instrument successfully isolated the monoisotopic peak from the complex isotopic envelope, which presumably contained ions derived from other biomolecules. Moreover, at such a high-resolution setting, we were able to identify the peptide as MBP (sequence position, 237–248) with a significant MASCOT score.

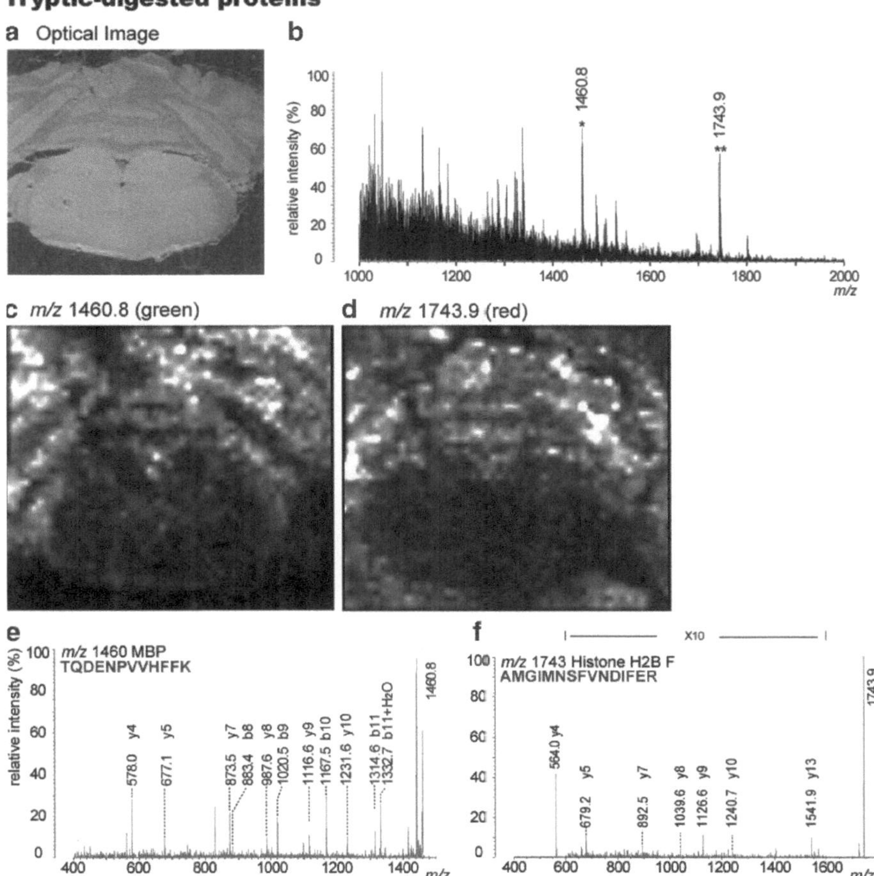

Fig. 16.5 Example of tryptic-digested protein imaging and precursor ion mass spectra with positive ion detection mode. (**a**) Optical image of imaging region. (**b**) Overall mass spectrum from imaging region. (**c**, **d**) Imaging results of m/z 1,460.8 and 1,743.9, which are labeled by *asterisks* in (**b**). These peaks were identified by direct MS^n and were identified as the fragment ions of myelin basic protein (MBP) (**f**) and histone H2B (**g**)

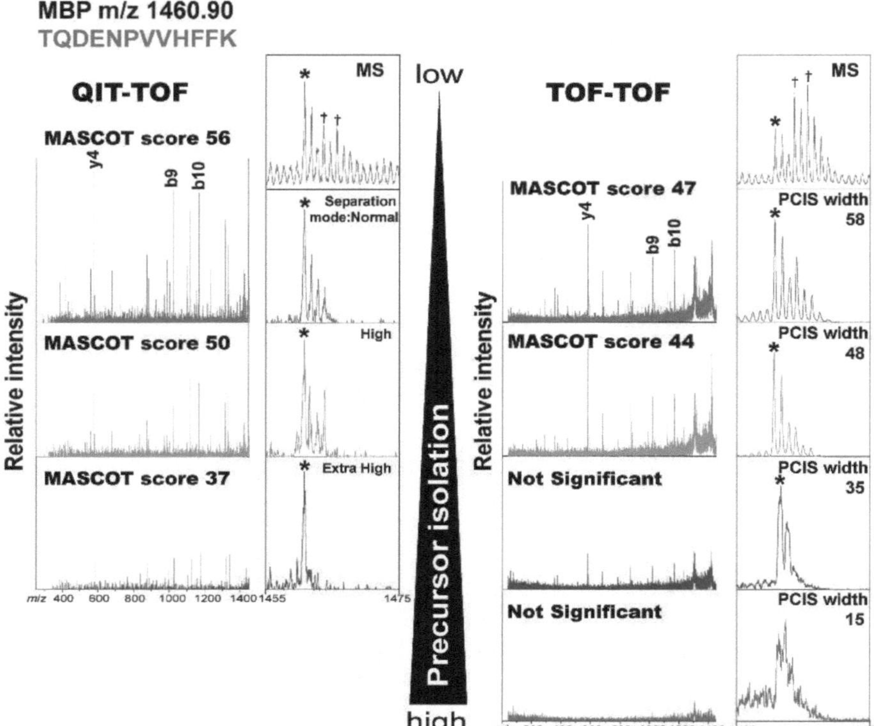

Fig. 16.6 Comparison of product ion spectrum of *m/z* 1,460.8 between QIT-TOF and TOF/TOF. Product ion spectra of (**a**) QIT-TOF and (**b**) TOF/TOF compared by changing the mass window of precursor ion selection. Comparison of (**c**) S/N and (**d**) mass resolution of fragment peaks between QIT-TOF and TOF/TOF (Reprinted from Shimma et al., Anal Chem 80(3):878–885.)

16.7 Conclusions

Two-dimensional imaging and molecular identification from tissue surfaces using AXIMA-QIT, which is a MALDI mass spectrometer with the QIT-TOF instrument, was presented. The uniqueness of the QIT instrument presented here enables us to perform molecular imaging with proper identification of interested molecules. As shown, the imaging system covers a wide range of application area in biological and clinical fields.

References

1. Chaurand P, Rahman MA, Hunt T, et al. (2008) Monitoring mouse prostate development by profiling and imaging mass spectrometry. Mol Cell Proteomics 7:411–423
2. Chaurand P, Cornett DS, Caprioli RM (2006) Molecular imaging of thin mammalian tissue sections by mass spectrometry. Curr Opin Biotechnol 17:431–436

3. Chaurand P, Schwartz SA, Caprioli RM (2004) Assessing protein patterns in disease using imaging mass spectrometry. J Proteome Res 3:245–252
4. Stoeckli M, Staab D, Staufenbiel M, et al. (2002) Molecular imaging of amyloid beta peptides in mouse brain sections using mass spectrometry. Anal Biochem 311:33–39
5. Jackson SN, Wang HY, Woods AS (2005) Direct profiling of lipid distribution in brain tissue using MALDI-TOFMS. Anal Chem 77:4523–4527
6. Sugiura Y, Konishi Y, Zaima N, et al. (2009) Visualization of the cell-selective distribution of PUFA-containing phosphatidylcholines in mouse brain by imaging mass spectrometry. J Lipid Res 50:1776–1788
7. Shimma S, Setou M (2007) Mass microscopy to reveal distinct localization of heme B (m/z 616) in colon cancer liver metastasis. J. Mass Spectrom Soc Jpn 55:145–148
8. Shimma S, Sugiura Y, Hayasaka T, et al. (2007) MALDI-based imaging mass spectrometry revealed abnormal distribution of phospholipids in colon cancer liver metastasis. J Chromatogr 855:98–103
9. Khatib-Shahidi S, Andersson M, Herman JL, et al. (2006) Direct molecular analysis of whole-body animal tissue sections by imaging MALDI mass spectrometry. Anal Chem 78:6448–6456
10. Martin RL, Brancia FL (2003) Analysis of high mass peptides using a novel matrix-assisted laser desorption/ionisation quadrupole ion trap time-of-flight mass spectrometer. Rapid Commun Mass Spectrom 17:1358–1365
11. Menon ST, Han M, Sakmar TP (2001) Rhodopsin: structural basis of molecular physiology. Physiol Rev 81:1659–1688
12. Hayasaka T, Goto-Inoue N, Sugiura Y, et al. (2008) Matrix-assisted laser desorption/ionization quadrupole ion trap time-of-flight (MALDI-QIT-TOF)-based imaging mass spectrometry reveals a layered distribution of phospholipid molecular species in the mouse retina. Rapid Commun Mass Spectrom 22:3415–3426
13. Kuerschner L, Ejsing CS, Ekroos K, et al. (2005) Polyene-lipids: a new tool to image lipids. Nat Methods 2:39–45
14. Emoto K, Kobayashi T, Yamaji A, et al. (1996) Redistribution of phosphatidylethanolamine at the cleavage furrow of dividing cells during cytokinesis. Proc Natl Acad Sci U S A 93:12867–12872
15. Ishitsuka R, Yamaji-Hasegawa A, Makino A, et al. (2004) A lipid-specific toxin reveals heterogeneity of sphingomyelin-containing membranes. Biophys J 86:296–307
16. Stark DD, Hendrick RE, Hahn PF, et al. (1987) Motion artifact reduction with fast spin-echo imaging. Radiology 164:183–191
17. Stark DD, Weissleder R, Elizondo G, et al. (1988) Superparamagnetic iron oxide: clinical application as a contrast agent for MR imaging of the liver. Radiology 168:297–301
18. Groseclose MR, Andersson M, Hardesty WM, et al. (2007) Identification of proteins directly from tissue: in situ tryptic digestions coupled with imaging mass spectrometry. J Mass Spectrom 42:254–262
19. Setou M, Hayasaka T, Shimma S, et al. (2008) Protein denaturation improves enzymatic digestion efficiency for direct tissue analysis using mass spectrometry. Appl Surf Sci 255(4): 1555–1559

Chapter 17
MALDI Imaging with Ion-Mobility MS: Waters Corporation

Motoji Oshikata, Yuki Sugiura, Naohiko Yokota, and Mitsutoshi Setou

Abstract For reasons of the unique detection principle based on mass spectrometry, MALDI-IMS provides an attractive proposition for molecular imaging, as a large range of different compounds can be detected simultaneously with simple protocol and fast data acquisition; also, it does not require any labeling of compounds of interest. In addition, the combination of ion-mobility separation with MALDI TOF-MS provides a unique separation dimension to further enhance IMS; ion-mobility spectroscopy can be used to produce images without interference from background ions of similar mass, which can remove ambiguity from imaging experiments and lead to more precise localization of the compound of interest, such as drugs and metabolites. Thus, ion-mobility TOF-MS has the potential to reduce complexity and improve confidence in imaging experiments.

17.1 Introduction

The past few years have seen a rapid increase in the interest in imaging mass spectrometry (IMS). The aim of this type of experiment is to obtain information on the spatial distribution of compounds in biological samples. Mass spectrometry (MS) is an attractive proposition for imaging, as a large range of different compounds can be detected simultaneously with simple protocol and fast data acquisition, and it does not require any labeling of compounds of interest.

M. Oshikata and N. Yokota
Waters Corporation, 1-3-12 Kita-shinagawa, Shinagawa-ku, Tokyo 140-0001, Japan

Y. Sugiura and M. Setou (✉)
Department of Bioscience and Biotechnology, Tokyo Institute of Technology,
4259 Nagatsuta-cho, Midori-ku, Yokohama, Kanagawa 226-8501, Japan

Department of Molecular Anatomy, Hamamatsu University School of Medicine,
1-20-1 Handayama, Higashi-ku, Hamamatsu, Shizuoka 431-3192, Japan
e-mail: setou@hama-med.ac.jp

M. Setou (ed.), *Imaging Mass Spectrometry*: *Protocols for Mass Microscopy*,
DOI 10.1007/978-4-431-09425-8_17, © Springer 2010

In earlier studies, IMS was developed as a tool for protein imaging [1–3], and in fact most of the reports concerning matrix-assisted laser desorption/ionization (MALDI)-IMS thus far describe the direct detection and imaging of proteins or peptides. On the other hand, research regarding the detection and imaging of small organic molecules has been increasing recently. The emergence of IMS as a tool for metabolite imaging has great impact because we do not have established technology for metabolite imaging whereas well-established methods have been widely applied for imaging of transcripts and proteins [4]. In this context, metabolic imaging by IMS is critical for interpretations of various aspects of life science.

Despite the promising capability of IMS, especially in the imaging of small metabolites, this emerging technique has several remaining problems. Below, we discuss point-to-point approaches to address these by using the MALDI-SYNAPT high definition MS (HDMS) system (Fig. 17.1).

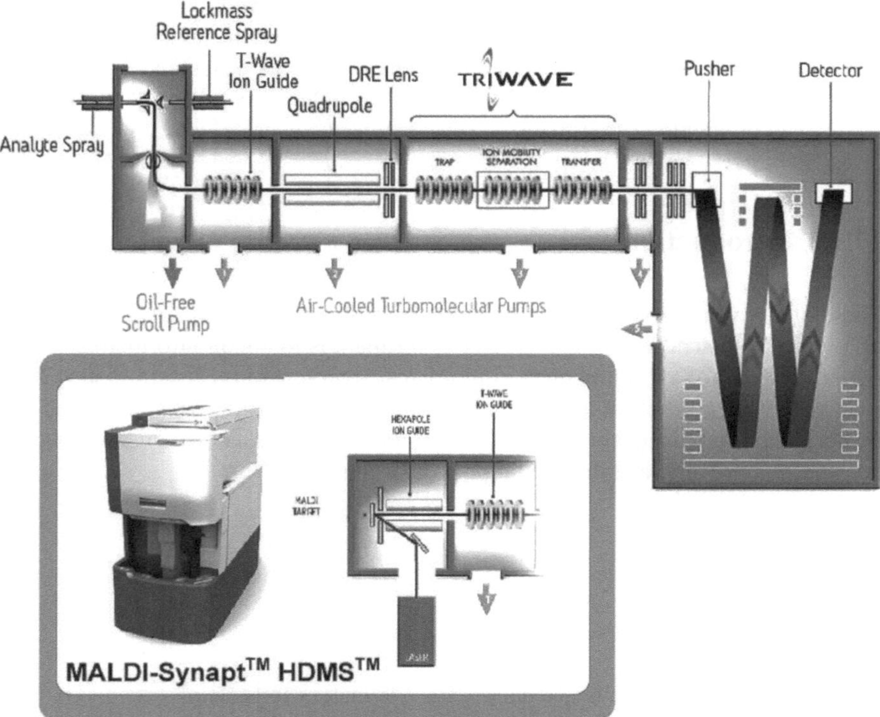

Fig. 17.1 Overview of MALDI-SYNAPT high definition (HD) MS: *Triwave* enables access to the unique benefits of ion-mobility spectroscopy, providing another separation capability; *oa-TOF* achieves high acquisition rates that complement ion-mobility separation time frames

17.1.1 Physical Properties of Tissue Samples

Tissue sections exhibit different physical properties from conventional MALDI samples: i.e., variation in electrical conductivity and sample morphology are observed [5].

In axial MALDI mass spectrometers, such as TOF/TOF systems, the source region, including the sample holder and sample, form part of the time-of-flight (TOF) system. Therefore, the properties of a tissue sample may result in a reduction of performance of the mass analyzer, leading to a decrease in mass accuracy and mass spectral resolution. This uncertainty carries the risk of incorrectly assigning masses. In this regard, the orthogonal acceleration mass analyzers used by MALDI-SYNAPT HDMS do not suffer from this limitation, as the source region does not form part of the mass analyzer and instrument performance is independent of the sample analyzed (see Fig. 17.1).

17.1.2 Sample Complexity

To produce unambiguous ion images, it is important to be able to separate and distinguish the compounds of interest from MALDI matrix background. In particular, the low-m/z region ($m/z < 1000$) of a MALDI spectrum contains a large population of ions from endogenous metabolites as well as matrix-related adduct clusters and fragments [6, 7]. Such a high density of ions means that multiple compounds often share the same nominal mass.

With the introduction of the MALDI-SYNAPT HDMS system, it is possible to separate ions using ion-mobility separation before mass analysis. Ion-mobility separation allows for the separation of ions according to their size, shape, and charge state. Using this technique it is possible to separate different compound classes, giving additional confidence that the true distribution of an ion of interest is observed, as discussed next.

17.1.3 Abundance of Compound of Interest

Compounds of interest may be difficult to detect when they are present at low concentration. This problem can be compounded by the sample complexity described previously.

Targeted MS/MS analysis is an effective solution; a substantial improvement in the signal-to-noise ratio can be achieved when monitoring specific fragment ions from compounds of interest in the MS/MS mode of analysis. A further advantage of this technique is that by monitoring more than one fragmentation, it is possible to validate the distribution of a compound of interest.

Below, we introduce an application example of MALDI-SYNAPT HDMS for imaging of small metabolites in a mouse spinal cord section. The example clearly demonstrates the usefulness of ion-mobility separation for small m/z compounds in MALDI-IMS.

17.2 Methods

- 2,5-Dihydroxy benzoic acid (DHB) matrix was used at a concentration of 50 mg/ml in 70% MetOH, 0.1% trifluoroacetic acid.
- The frozen spinal cord sections were thaw-mounted on indium tin oxide-coated glass slides.
- The sections were coated with matrix using an airbrush (Procon Boy FWA Platinum; Mr. Hobby, Tokyo, Japan).
- The sample was analyzed using a MALDI-SYNAPT HDMS system.
- The instrument was operated in V-mode over a m/z range of 50–1,000 in positive ion detection mode.
- Spatial resolution at 200 μm was selected, and 200 raster shots were acquired per pixel at a laser repetition rate of 200 Hz.
- The area to be imaged was selected using MALDI Imaging Pattern Creator (Waters Corporation, Manchester, UK).
- After acquisition, HDMS data evaluation was performed using DriftScope (Waters Corporation, Manchester, UK).
- Data were converted into Anatyze file format using MALDI-Imaging Converter (Waters Corporation, Manchester, UK) for image analysis using BioMap (Novartis, Basel CH).

Supported by a number of software tools, the MALDI imaging workflow on the MALDI-SYNAPT HDMS system has been optimized to make the production of high definition imaging data as easy as possible. The steps involved in this workflow are outlined in Fig. 17.2.

17.2.1 Selection of Area for Mass Analysis

It is important to limit the area to be analyzed to the tissue of interest, as this saves a great deal of instrument time, especially for high-resolution images. The Creator imaging software simplifies the process, using a digital image of the prepared sample to guide selection of the areas of interest. Area selections are made using a drawing tool, allowing rectangular, elliptical, and free-drawn areas to be selected. Multiple areas can be selected for one experiment. The laser coordinates corresponding to the selected areas are automatically determined at a user-definable spatial resolution and then used during data acquisition.

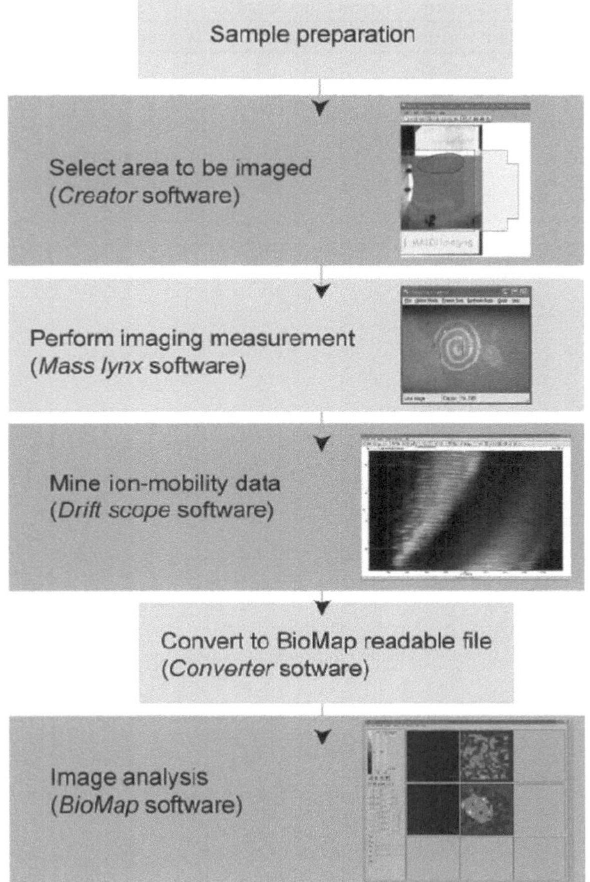

Fig. 17.2 Workflow of MALDI ion-mobility IMS with SYNAPT HDMS

17.2.2   Automated HDMS or MS Imaging Experiments

Imaging data acquisition is managed from the MassLynx software sample list. Researchers can choose between MS, MS/MS, HDMS, and HDMS/MS imaging experiments. Details of these experiments are shown in Table 17.1.

17.2.3   Ion-Mobility Data Conversion to Imaging Format

After data acquisition in HDMS or HDMS/MS mode, the mobility data can be viewed and mined using DriftScope software (see Fig. 17.3). A specific m/z and drift-time range can be selected and converted into ion intensity maps. These intensity

Table 17.1 MALDI-IMS experiments available on the MALDI-SYNAPT HDMS system

Type of imaging experiment	Ion-mobility separation	Type of mass spectral data
MS	No	MS
MS/MS	No	MS/MS on one or more precursor ions
HDMS	Yes	MS
HDMS/MS	Yes	Trap fragmentation
		Transfer fragmentation
		TAP fragmentation

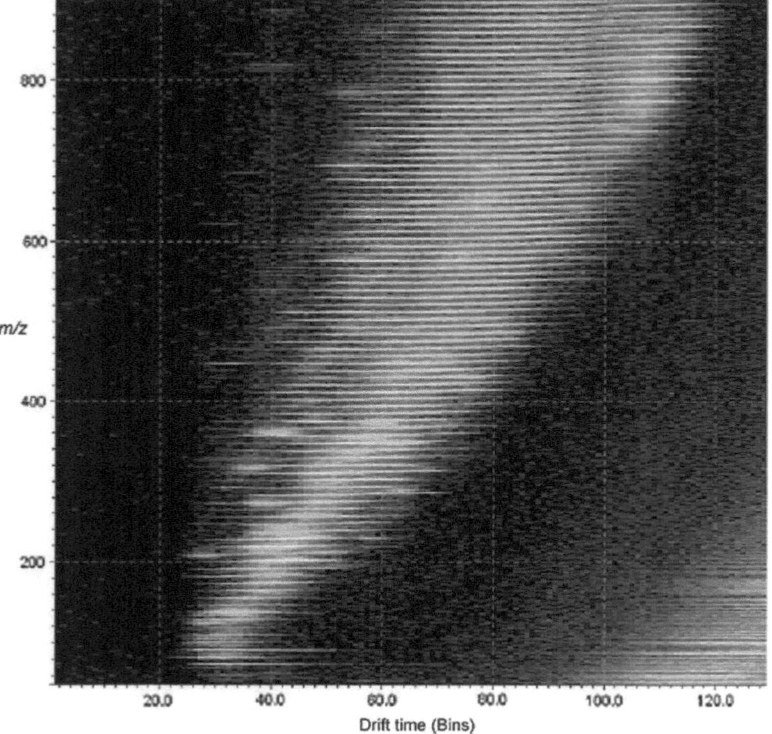

Fig. 17.3 MALDI ion-mobility 2D plot (*m/z* vs. drift time) of a mouse spinal cord tissue section with 2,5-dihydroxy benzoic acid (DHB) matrix in positive ion mode

maps can be imported into BioMap (Novartis, CH) using the MALDI imaging converter software tool.

17.3   Results and Discussion

Owing to the complexity of the tissue samples under investigation, a major limitation of MALDI-IMS is a risk of isobaric ions distorting the ion distribution and thus invalidating results, especially when applying single-stage MS imaging measurement.

In this regard, the MALDI-SYNAPT HDMS system achieves this by gas-phase separation of ions based on their size, shape, and charge using high-efficiency ion mobility. In the imaging mode, a full three-dimensional data set is acquired at every spatial position consisting of m/z, drift time, and ion intensity, ensuring maximum flexibility for mining the data.

Figure 17.3 shows a typical MALDI ion-mobility TOF spectrum (plot of m/z, y-axis vs. drift time, x-axis), obtained from the mouse spinal cord section. In a previous study using standard compounds it was demonstrated that different classes of biomolecules such as lipids, oligonucleotides, and peptides form distinct familial trend lines on a two-dimensional (2D) plot of mobility drift times versus m/z [8]. Therefore, we can mine a highly specific imaging dataset for molecular species of interest by ion-mobility separation [9].

In the drift-time spectrum, an independent peak was observed around 110 bins (arrowhead in Fig. 17.4a). It has been demonstrated that lipid ion species were consistently slower than isobaric peptides and DHB matrix clusters [8, 9]. Based on previous studies [9, 10], we assumed the drift-time peak as a lipid-derived signal. Then, we extracted mass spectra from a distinct lipid trend area that is separated

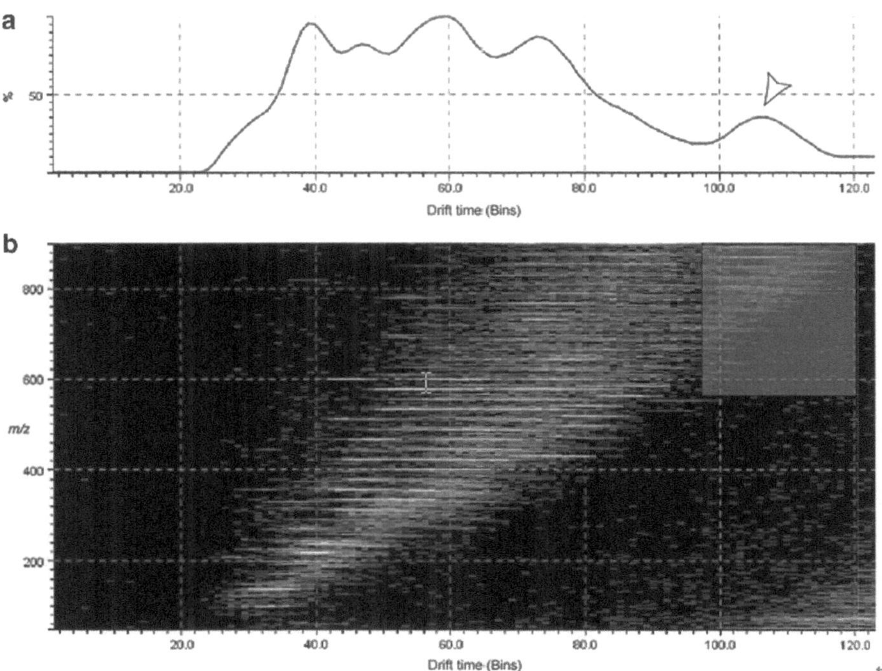

Fig. 17.4 Selection and separation of lipid-related signals by difference of drift time. (**a**) Overall drift time spectrum. *Arrowhead* represents a peak corresponding to lipid species [9]. (**b**) MALDI ion-mobility 2D plot (*m/z* vs. drift time) of a mouse spinal cord tissue section with DHB matrix in positive ion mode. A *square region* was selected (98–230 bins and 580–900 *m/z*), and using the 2D range of mobility time and mass, mass spectra were reconstructed for each spatial point

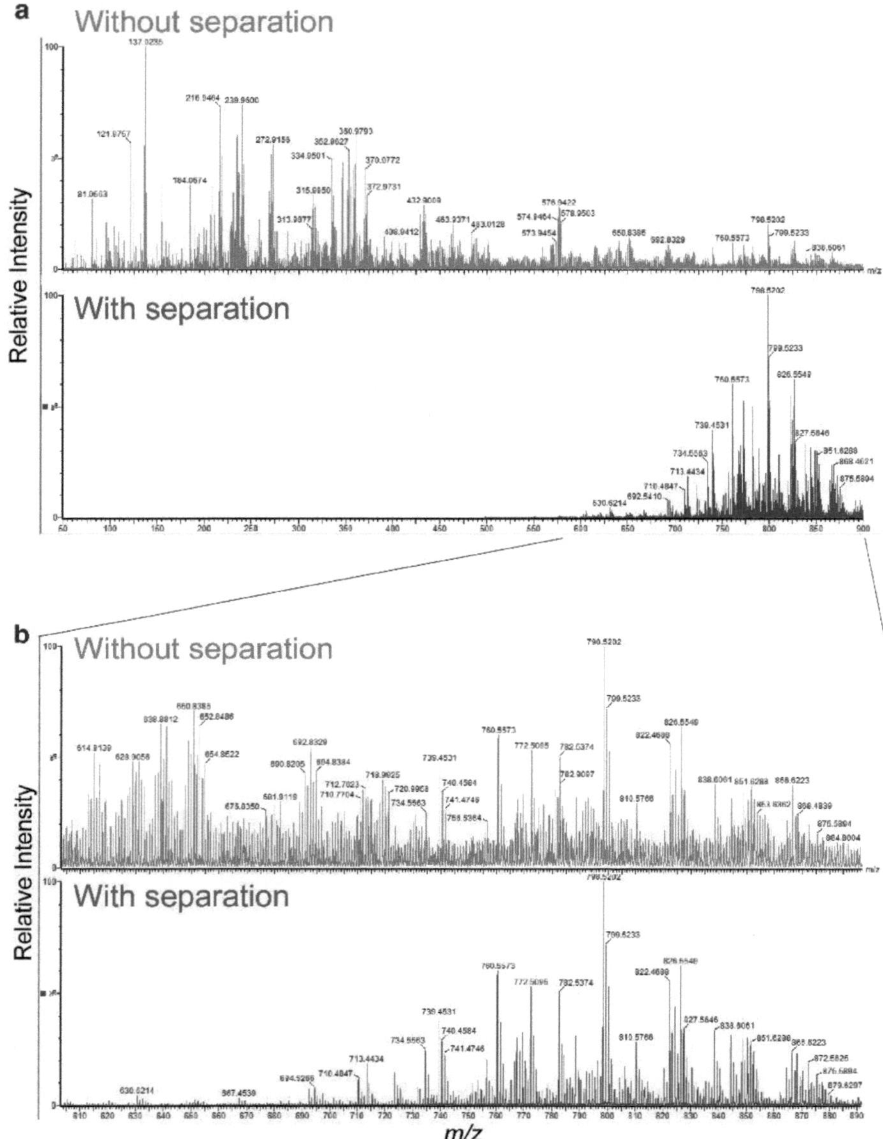

Fig. 17.5 Ion-mobility separation results in specific MS profiles for lipid-related signals

from other biomolecules (square region in Fig. 17.4b). It is this 2D range of mobility time and mass that was used to acquire the lipid images in the following study.

In Fig. 17.5, which shows mass spectra with/without ion-mobility separation, it can clearly be seen that ions of lipid species were separated by the ion-mobility

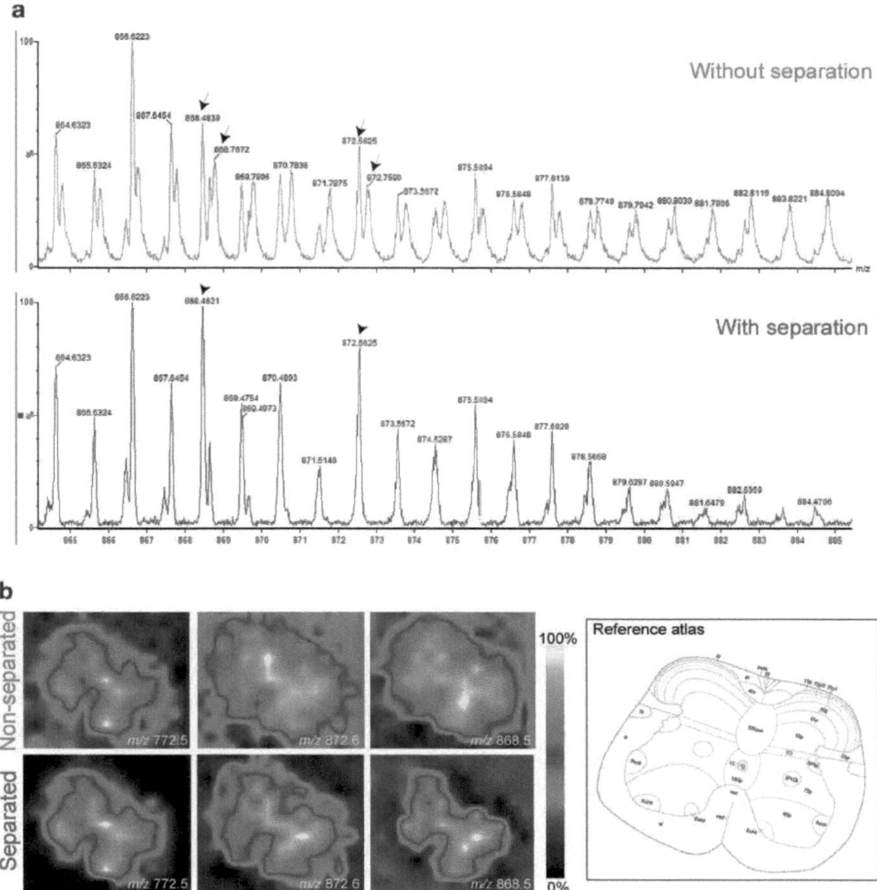

Fig. 17.6 Ion-mobility separation helps to provide the true spatial tissue distribution of the endogenous metabolite by removing any contribution of the interfering matrix ions. (**a**) Mass spectrum with/without ion-mobility separation. Signals presumably derived from matrix clusters were removed by the separation (*arrows*), and signal-to-noise ratio of lipid signals was enhanced (*arrowheads*). (**b**) Ion distribution images of phosphatidylcholines reconstructed from mass spectra with or without ion-mobility separation

dimension from the complex ion mixture containing other trends of ion molecules, such as DHB clusters.

Figure 17.6 demonstrates that ion-mobility separation provides a more specific ion distribution image of compounds by removing matrix-related signals. Mass peaks presumably derived from matrix clusters were removed by the separation process (arrows in Fig. 17.6a). On the other hand, the signal-to-noise ratio of lipid signals was improved (arrowheads).

Such improvement on signal selectivity clearly helps to provide the true spatial tissue distribution of the endogenous metabolite by removing any contribution of the interfering matrix ions. Fig. 17.6b shows ion images for the ion of phosphatidylcholines with or without ion-mobility separation [11]. When ion-mobility separation is applied, the ion images for phosphatidylcholines have a significantly lower background noise level, as was especially true for the lipids at low concentration (m/z 872.6 and 868.5).

17.4   Conclusions

- The combination of ion-mobility separation with MALDI provides a unique separation dimension to further enhance IMS.
- Ion-mobility spectroscopy can be used to produce images without interference from background ions of similar mass; this can remove ambiguity from imaging experiments and lead to more precise localization of the compound of interest, such as drugs and metabolites.
- HDMS has the potential to reduce the complexity and improve confidence in imaging experiments.
- MALDI imaging offers a complementary approach to established imaging techniques, such as whole-body autoradiography. It has the advantage that no labeling of the compounds of interest is required and that multiple compounds can be monitored simultaneously, e.g., drugs and metabolites. As a result, the experiments can take place rapidly and the technique is relatively inexpensive.

References

1. Caprioli RM, Farmer TB, Gile J (1997) Molecular imaging of biological samples: localization of peptides and proteins using MALDI-TOF MS. Anal Chem 69:4751–4760
2. Chaurand P, Stoeckli M, Caprioli RM (1999) Direct profiling of proteins in biological tissue sections by MALDI mass spectrometry. Anal Chem 71:5263–5270
3. Stoeckli M, Chaurand P, Hallahan DE, et al. (2001) Imaging mass spectrometry: a new technology for the analysis of protein expression in mammalian tissues. Nat Med 7:493–496
4. Sugiura Y, Setou M (2009) Current imaging mass spectrometry for small metabolites. J Mass Spectrom Soc Jpn (in press)
5. Sugiura Y, Shimma S, Setou M (2006) Two-step matrix application technique to improve ionization efficiency for matrix-assisted laser desorption/ionization in imaging mass spectrometry. Anal Chem 78:8227–8235
6. Garrett TJ, Prieto-Conaway MC, Kovtoun V, et al. (2006) Imaging of small molecules in tissue sections with a new intermediate-pressure MALDI linear ion trap mass spectrometer. Int J Mass Spectrom 260:11
7. Cornett DS, Frappier SL, Caprioli RM (2008) MALDI-FTICR imaging mass spectrometry of drugs and metabolites in tissue. Anal Chem 80:5648–5653
8. Woods AS, Ugarov M, Egan T, et al. (2004) Lipid/peptide/nucleotide separation with MALDI-ion mobility-TOF MS. Anal Chem 76:2187–2195

  9. Jackson SN, Ugarov M, Egan T, et al. (2007) MALDI-ion mobility-TOF MS imaging of lipids in rat brain tissue. J Mass Spectrom 42:1093–1098
10. McLean JA, Ridenour WB, Caprioli RM (2007) Profiling and imaging of tissues by imaging ion mobility-mass spectrometry. J Mass Spectrom 42:1099–1105
11. Sugiura Y, Konishi Y, Zaima N, et al. (2009) Visualization of the cell-selective distribution of PUFA-containing phosphatidylcholines in mouse brain by imaging mass spectrometry. J Lipid Res (in press)

Part VIII
Comparison of SIMS and MALDI for Mass Spectrometric Imaging

Chapter 18
Comparison of SIMS and MALDI for Mass Spectrometric Imaging

John S. Hammond

Abstract Secondary ion mass spectrometry (SIMS) has become a highly refined and widely used mass spectrometric technique for the two- and three-dimensional (3D) imaging of inorganic materials, especially for the semiconductor industry. The optimization of primary ion sources for ultrahigh spatial resolution, very high transmission and high mass resolution analyzers, and specialized sample handling, particularly for wafers, has resulted in the widespread utilization of SIMS instruments for materials research and quality control in the semiconductor and magnetic materials industries. The utilization of SIMS for mass spectrometric imaging of organic and biological samples has trailed the applications for inorganic materials because of the need to improve the sub-micrometer molecular spatial resolution capabilities of SIMS that are now commonplace for inorganic and semiconductor SIMS analysis. The rapid improvements in matrix-assisted laser desorption/ionization (MALDI) for imaging macromolecular species, particularly for tissue cross sections, are now being limited by the fundamental spatial resolution capabilities of the laser desorption process. A comparison of the new developments of SIMS and MALDI instrumentation points out that these two techniques are complementary in many capabilities, especially ultimate spatial resolution and molecular mass range. An overview is presented of the SIMS process, primary ion sources, mass analyzers, and software for mass spectrometric imaging of organic and especially biological samples. Examples of the analysis of tissue cross sections, imaging resolution of subcellular features, and 3D imaging of drug delivery materials illustrate the potential for SIMS and MALDI to become invaluable complementary techniques for future mass spectrometric imaging.

J.S. Hammond (✉)
Physical Electronics, 18725 Lake Drive East, Chanhasse, MN 55317, USA
e-mail: JHammond@phi.com

M. Setou (ed.), *Imaging Mass Spectrometry: Protocols for Mass Microscopy*,
DOI 10.1007/978-4-431-09425-8_18, © Springer 2010

18.1   Introduction

Secondary ion mass spectrometry (SIMS) has become a highly refined and widely used MS technique for the two-dimensional (2D) and three-dimensional (3D) imaging of inorganic materials, especially in the semiconductor industry. The optimization of primary ion sources for ultrahigh spatial resolution, very high transmission and high mass resolution analyzers, and specialized sample handling – particularly for wafers – has resulted in the widespread utilization of SIMS instruments for materials research and quality control in the semiconductor and magnetic materials industries. The use of SIMS for the MS imaging of organic and biological samples has trailed applications for inorganic materials because of the need to improve the submicrometer molecular spatial resolution capabilities of SIMS – capabilities that are now commonplace in inorganic and semiconductor SIMS analysis. Rapid improvements in matrix-assisted laser desorption/ionization (MALDI) for imaging macromolecular species, particularly for tissue cross sections, are now being limited by the fundamental spatial resolution capabilities of the laser desorption process. A comparison of new developments in SIMS and MALDI instrumentation indicates that these two techniques are complementary in many ways, especially with respect to the ultimate spatial resolution and molecular mass range. An overview is provided here of the SIMS process, primary ion sources, mass analyzers, and the software used for the MS imaging of organic and especially biological samples. Examples of the analysis of tissue cross sections, the imaging resolution of subcellular features, and the 3D imaging of drug delivery materials illustrate the potential for SIMS and MALDI to become invaluable complementary techniques in undertaking future MS imaging procedures.

18.2   SIMS Overview

18.2.1   Analytical Overview

In SIMS, a primary ion beam is directed to the sample surface to induce a collision cascade in the near-surface region. A portion of the energy from this collision cascade is redirected toward the surface, thus promoting the release of atoms, molecules, and molecular fragments into the vacuum system of the SIMS instrument. This release of surface material is commonly referred to as the sputtering process. A fraction of the sputtered material emerges from the sample as positively or negatively charged species; these are then extracted into a mass spectrometer, where ions are separated on the basis of their respective mass-to-charge ratios.

The ejected ions are collected by a high-performance mass spectrometer. Today, two types of spectrometers are used in MS imaging: a time-of-flight (TOF) analyzer and a magnetic sector analyzer. Each of the current analyzer designs collects about 50% of the emitted ions and can provide a mass analysis with an ultrahigh mass resolution [i.e., >15,000 M/Δm full-width at half-maximum (FWHM)]. The magnetic sector analyzer separates and counts, in a parallel fashion, a limited number of different mass-to-charge species, depending on the number of detectors in the detection system. The TOF analyzer, coupled with a pulsed primary ion gun, separates all the ions as a function of their flight time; it can therefore collect the full mass range of emitted secondary ions.

The imaging capabilities of current commercial SIMS instruments is based on the use of scanning microprobe primary ion sources, with beam diameters as small as 50 nm. The use of scanning primary ion beams that are completely computer controlled and synchronized with a TOF or magnetic sector mass spectrometer allows for the collection of mass spectra or ion-selected images over the field of view defined by the user. The advantage of the TOF-SIMS design is the parallel collection of ion images from all the different ion species, although the ultimate spatial resolution and analytical speed for a single mass-selected ion image does not equal the ultimate spatial resolution performance of the magnetic sector-designed SIMS instruments. Commercially available SIMS instruments include software programs that provide enhanced image presentations and a range of image analysis capabilities.

It is beyond the scope of this text to provide a comprehensive discussion of all aspects of the historical development, instrumentation engineering details, and sophisticated data interpretation available for SIMS analysis. For more detail, the reader is referred to two of several available treatises on this subject [1, 2].

18.2.2 Primary Ion Sources and Ion Creation

Figure 18.1 shows a schematic representation, produced by molecular dynamics simulation calculations, of the impact of a 15-keV Ga primary ion into an Ag-111 crystal, as well as the resultant energy cascade [3].

This simulation illustrates several key points concerning the SIMS process. First, the primary ion releases energy below the surface of the sample in the energy cascade. This released energy can result in significant damage to chemical bonds and the molecular structure. A portion of the energy cascade reaches the sample surface and initiates the sputtering process. For this reason, initial SIMS experiments that aimed to obtain molecular spectroscopy and molecular imaging without chemical or molecular damage limited the primary ion dose to 1% of the surface atoms, or 10^{13} ions/cm^2. This was defined as the static limit for static SIMS (SSIMS), and it was intended to ensure that no primary ion struck a point on the sample surface that had already sustained a previous ion impact. Second, the number of sputtered atoms, which is dependent on the primary ion mass and kinetic energy and the composition of the sample, is relatively low. In the foregoing simulation, only 29 atoms were sputtered for each incoming primary ion. In addition, the probability of atoms or molecular fragments leaving the sample surface as ions is also dependent on the chemical composition of the sample. For some inorganic compounds,

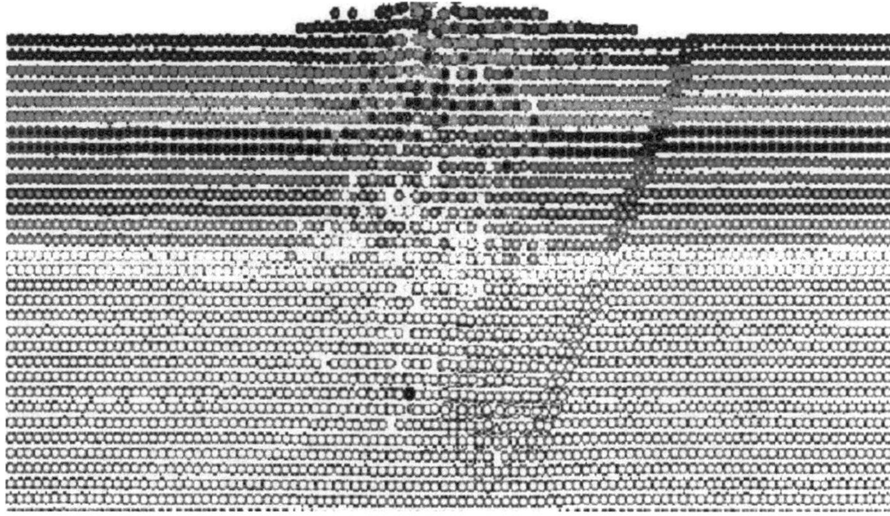

Fig. 18.1 Molecular dynamics simulation of impact of a 15-kV Ga ion into Ag, after 29 ps

the probability of ion formation during sputtering can approach unity; for many molecular species, especially large organic species, the probability can be as low as 10^{-3}. The variability of the probability of ion formation with changes in surface composition has led to the working premise that SIMS is a nonquantitative technique and that changes in intensity among SIMS images could result from both the change in ion formation probability and the surface coverage. Its low ion formation probability has led many researchers to suggest that imaging SSIMS for organic and biological samples may be limited to a spatial resolution of 1 μm.

The separation of the primary ion point of impact and the point of secondary ion emission is suggested in the molecular dynamics simulations of Fig. 18.1 and further illustrated in Fig. 18.2.

This illustration has been explored in greater analytical detail by several research groups, including a recent publication by Delcorte and Garrison [4].

The coincidence within 10 nm of the primary ion impact and large mass molecular ion fragment emission points to the potential for SIMS to provide excellent spatial resolution for MS imaging. For this reason, early SIMS instruments sought to use highly focused scanning primary ion sources. Although a highly focused Cs ion beam source with a minimum beam diameter <50 nm is used with the Cameca nanoSIMS 50 instrument [5], most SIMS instruments use a liquid metal ion gun (LMIG) as the source for primary ions. The Cs source in the Cameca nanoSIMS 50 operates with a rastered DC beam, which can provide outstanding spatial resolution for atomic secondary ions but is generally not useful for organic or molecular species; for this reason, the nanoSIMS 50 is not discussed further in this chapter.

The early LMIG sources for SIMS instruments all used a Ga ion emitter. When used in conjunction with a TOF analyzer, the LMIG must be pulsed at each pixel

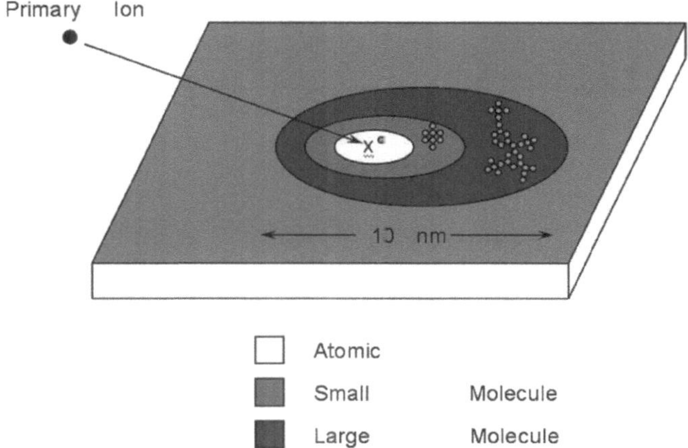

Fig. 18.2 Sites for atomic, small, and large molecular ion ejections, relative to primary impact site

of the scanned image over the sample surface. The time-duration of this pulse is generally <1 ns for high mass resolution spectroscopic data. In the imaging mode, the Ga-emitting LMIG sources can routinely produce a pulsed beam diameter <100 nm, with slightly degraded mass resolution performance. This performance is highly useful for most inorganic and semiconductor samples. However, because of the low sputter yield and ionization efficiency of Ga primary ion sources, the practical limit for MS imaging on organic and biological samples has been ~1 μm.

To improve both the higher mass fragment sputter yield efficiency by the primary ion and the probability of ionization in the sputtered species, both of which are important for improving the ion-counting statistics necessary for the imaging spatial resolution of organic and biological samples, alternative primary ion sources are being explored. A summary of the presently available micro-focused pulsed primary ion sources for SIMS imaging is shown in Table 18.1. The relative ion yield efficiency is based on the relative number of high mass secondary ions (mass of m/z 300–400) of species from thick samples of organic films. The same number of primary ions was used for all measurements. The relative ion yields and the efficiency of secondary ion formation are important for the optimization of imaging SIMS experiments. It should be noted that the absolute numbers shown in Table 18-1 can change by about 50%, depending on the sample used. The key trends are that the relative ion yield is a nonlinear function of the mass of an atomic ion, and the use of a higher number of atoms in a cluster ion source also produces a dramatic nonlinear increase. To date, a C_{60}^+ source has been found to produce the highest relative ion yield for species from most thick organic materials, but the spatial resolution in a scanning pulsed microprobe mode is still limited to about 1 μm. The relative ion yield of the C_{60}^+ primary ion source is sufficiently high that the ion imaging spatial resolution may approach 100 nm, if improved primary ion source optics or new imaging analyzers can be developed.

Table 18.1 Performance of pulsed imaging ion sources [6]

Primary ion species	Ion mass a.m.u.	Relative ion yield efficiency	Best pulsed spatial resolution (μm)
Ga^+	69	1	80
In^+	115	3.5	100
Au^+	197	22	100
Bi^+	208	n.a.	100
Au_2^+	394	190	130
Bi_2^+	416	n.a.	n.a.
Au_3^+	591	270	130
Bi_3^+	624	n.a.	100
C_{60}^+	720	520	1,200

n.a. not applicable

18.2.3 *SIMS Analyzers*

Presently, two commercial designs are used for the TOF-SIMS MS imaging of organic and biological samples. The first is a reflectron design, which is second-generation TOF optics that was introduced after the initial introduction of TOF-SIMS instruments which used the Poschenrieder mass analyzer design [2]. The reflectron design has been refined in several models and is presently commercially available as the TOF.SIMS 5 from ION-TOF GmbH [7]. A schematic of this instrument is shown in Fig. 18.3.

The third-generation TOF-SIMS instrument is based on a triple focusing time-of-flight (TRIFT) mass analyzer [8]. This instrument is commercially available from Physical Electronics USA and ULVAC-PHI. A schematic diagram of this instrument is shown in Fig. 18.4.

Both analyzers provide similar high mass resolution and high collection efficiency for a broad mass range of secondary ions. In each instrument, the secondary ions are generated by a short time-duration pulse of highly focused primary ions. When they leave the sample surface, the ions have a spread of kinetic energy and angular distribution as a result of the primary ion energy cascade. To enhance the collection efficiency of the analyzer, the ions are focused through a high-voltage extraction field into the first lens of the analyzer. Due to the short time pulse of the primary ion beam, all mass ions leaving the sample are assumed to depart at $T=0$. All secondary ions are accelerated through the same voltage, such that the extraction voltage equals $0.5 \, mv^2$. The mass-to-charge ratio of each secondary ion will then be proportional to the square of the flight time to the detector. By using a series of low-mass/charge ions that appear in all spectra, the recorded flight times for all ions can be converted to exact masses. In both analyzers, the range of kinetic energy levels of the emitted secondary ions is minimized by the energy compensation trajectories inside the analyzer. The reflectron uses an ion mirror that allows ions with slightly higher kinetic energy levels to have longer flight paths into the ion mirror before their flight trajectories are "reversed" back to the detector. In the TRIFT

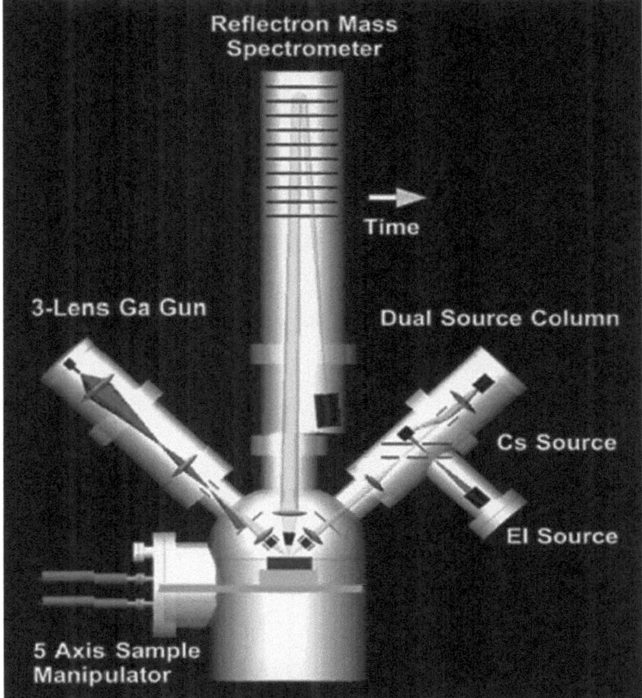

Fig. 18.3 Schematic of the TOF.SIMS 5

analyzer, secondary ions with higher kinetic energy levels traverse slightly larger radius trajectories, through the three 90° electrostatic sectors, before reaching the detector. By using either of these energy compensation techniques, the ions with the same m/z that start to diverge at the start of the flight path end up converging at exactly the same flight time at the detector; this allows for a very high mass resolution. By using a detector with a typical 20-kV value post acceleration, either of these two analyzers can achieve a mass resolution >15,000 M/Δm FWHM, with a mass range over m/z 10,000.

18.2.4   Two-Dimensional Analyses

As previously described in Sects. 18.2.1 and 18.2.2 (above), the pulsed microprobe ion beam can be scanned at the surface of the sample and the secondary ions of all masses are collected in a parallel manner by the TOF analyzer. The computer data system therefore records a matrix of the detected ions, with each detected ion labeled in the x- and y-positions of the primary ion beam and the flight time, t, of the ion to the detector. The calibration of the analyzer allows the time, t, to be

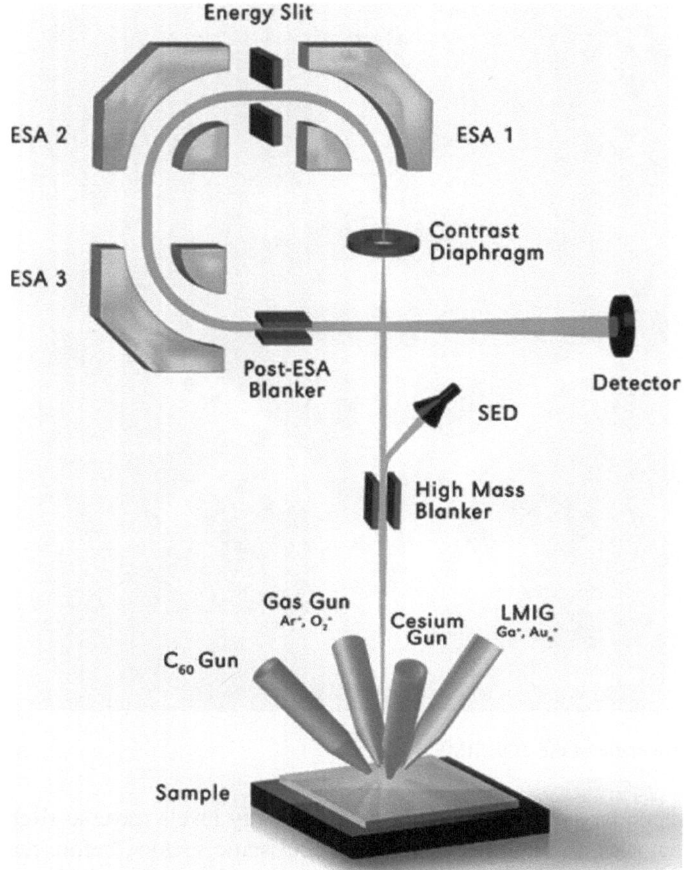

Fig. 18.4 Schematic of the PHI TRIFT V *nanoTOF* TOF-SIMS

converted into the mass of the detected ion. This matrix of data is referred to as the "raw file," and it is used for retrospective 2D image analyses. Depending on the optical constraints of the primary ion gun, the size and number of the x and y pixels in the image, and the desired mass range – and therefore the pulsed repetition rate of the analyzer – a SSIMS 2D raw file usually requires 15–30 min to acquire data for a typical $100\,\mu m \times 100\,\mu m$ surface image.

The flow process of the retrospective data analysis of the raw file is illustrated in Fig. 18.5. The user can select four different modes of data interpretation, each of which is briefly described next. These four forms of output are (1) total ion image, (2) total area spectrum, (3) regions of interest spectra, and (4) mass-specific images.

For MS imaging, probably the most important analytical question is how many different regions exist on the sample of interest. This can be answered, to a first approximation, by displaying the total ion image. For the total ion image, the software

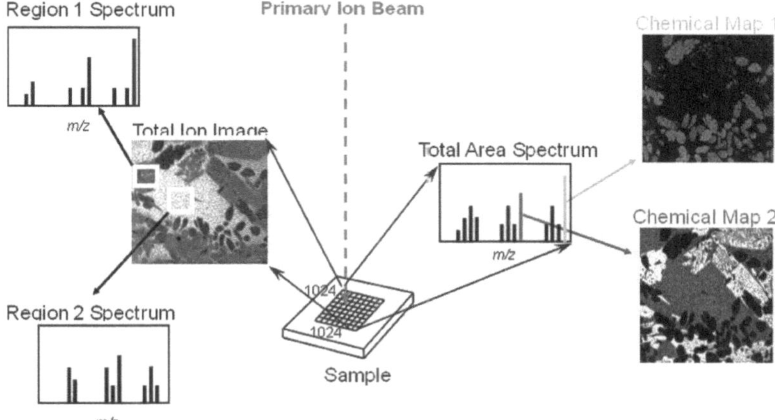

Fig. 18.5 Flow process for two-dimensional (2D) image analysis, using retrospective data analysis of a raw file

sums the number of ions, regardless of mass-to-charge ratio, at each pixel in the image. Normally, the different total ion intensity levels are displayed using a pseudo-color histogram routine; the areas of the image with the same color may then be assumed to be areas with similar chemical or molecular compositions. The total ion image can then also act as a "roadmap" for the locations or "regions" that may be targets for additional MS interrogation.

The second question is how many different chemical and molecular species exist within the imaged area. The software will sum all the ions of the same mass, regardless of x- and y-positions, from the raw file, thus producing a total area spectrum. The peaks in this spectrum can then be assigned by using exact mass assignments, by using comparisons with mass fragment libraries and standard molecular spectra that are available within the commercial TOF-SIMS software, or by using search engines in the system software to compare the total mass spectrum with reference spectra available from third-party sources (e.g., SurfaceSpectra [9]).

The third question is to identify the chemical and molecular compositions within specific regions of interest, identified using the total ion image. The location of the desired pixels located within a region can be selected manually by the user, using the total ion image as a roadmap, or the software can use the intensity values of the pixels for the histogram threshold selection of pixel locations. By summing the spectra from multiple pixels with the same range of total ion intensity over the original total ion image, excellent spectra can usually be reported from each region of interest. These spectra can then provide classifications for all regions of interest.

The final question is about obtaining images for each of the selected chemical or molecular species. One single peak – or a "fingerprint" of multiple peaks associated with one particular species – can be used to create a mass-selected image. The number of mass-selected images can then equal the number of species on the surface of the sample.

18.2.5   Depth Profile Analysis and 3D Analyses

As noted in the earlier section on ion creation (Sect. 18.2.2), the TOF-SIMS technique measures ions that usually originate from the outermost one to two atomic or molecular layers. Although the energy cascade process can disturb the chemical and molecular bonds below the surface in most organic and biological samples, many samples of metal, semiconductor, ceramic, and other inorganic materials can retain their elemental and chemical compositions, despite the affects of the energy cascade introduced by the primary ion beam. In the TOF-SIMS experiment, the short time-duration of the pulsed ion beam sputter removes a fraction of a monolayer during the typical MS analysis. However, for many inorganic samples, the experimenter is interested in gradients of composition, from the outermost surface to microns in depth. In addition, by combining the 2D imaging discussed in the previous section with an effective method of profiling the sample as a function of depth, a method of rendering 3D characterizations can be obtained. To facilitate a reasonable experimental time-duration and maintain high spatial resolution, high mass resolution, and high sensitivity in the TOF-SIMS analysis, the dual-beam approach to depth profiling is normally utilized. In this approach, one ion beam is used to produce secondary ions for SIMS analysis, and a second ion beam is provided for rapid sputter removal of the sample surface. This approach is illustrated in Fig. 18.6.

In the dual-beam approach, the MS analysis of the surface is based on the pulsed primary ion beam. Typically, the pulsed primary ion beam is rastered over a region of the sample, to acquire a raw file of the first layer. The computer control of the system then suspends the TOF-SIMS data acquisition and initiates a second DC

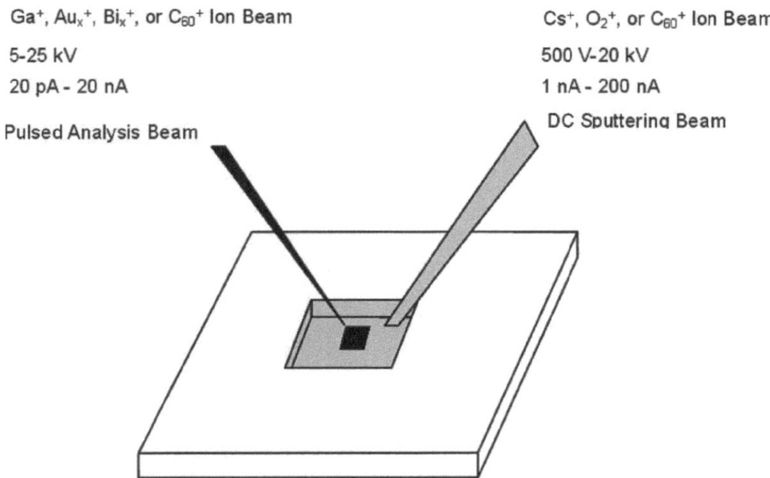

Fig. 18.6 Schematic of the dual ion beam approach to TOF-SIMS depth profiling and 3D analyses

beam to be rastered over a slightly larger area of the sample, to remove by sputtering a user-defined thin layer of the sample. After this layer is removed, the second ion beam is suspended and the pulsed primary ion beam is again used to acquire a second image raw file of freshly exposed surface. This process is alternately continued until a sputter depth profile to the desired depth has been obtained.

One significant advantage of the dual ion beam approach is that the composition and kinetic energy of the DC sputter beam can be selected to achieve a high sputter rate, with a minimum penetration depth of the energy cascade. In general, the lower the impact voltage of the DC sputtering ion beam, the lower will be the penetration depth of the energy cascade; this minimizes artifact formation within the composition of the sputter-exposed surface layers. In addition, the probe size and ion current of the sputter ion beam are both kept relatively large to promote rapid depth profiling; meanwhile, the probe size and ion current of the primary ion beam are kept small, to obtain optimal spatial resolution and optimal MS performance.

The depth profiling of organic and biological materials, in contrast to the depth profiling of inorganic materials, has not been successful when using monatomic ions. This failure has generally been attributed to the destruction of the chemical and molecular structure within the sample by the penetration of the energy cascade below the surface of the sample, as noted in Fig. 18.1. In the pioneering work of Gillen et al. in 1998 at NIST, the concept of organic and molecular depth profiling using cluster ion sources was demonstrated using polyatomic SF_5^+ and C_n^- primary ion sources [10]. Although sputter yields, secondary ion yields, and the potential for organic and molecular depth profiling have been studied for a large number of different polyatomic and cluster ion sources [11], C_{60}^+ is the most commonly used primary cluster ion source in use today, which is the result of its relatively high secondary ion yield and its commercial availability as a robust source; it also has favorable scanning microprobe and pulsed source capabilities [12].

The molecular dynamics simulations by Garrison et al. of the impact of a C_{60}^+ primary ion on a model Ag substrate, as shown in Fig. 18.7, illustrates key points vis-à-vis the success of molecular depth profiling [3]. The first point is the very large number of sputtered Ag atoms from the impact of one C_{60}^+ ion. In comparison with 21 atoms sputtered with the 15-keV Ga^+ primary ion, the 15-keV C_{60}^+ primary ion sputters 324 atoms. This sputter rate enhancement is even more pronounced when using one of many polymer systems, where the sputter rate of most polymers is eight times the sputter rate of SiO_2 under the same C_{60}^+ sputtering conditions [13]. This extremely high sputter rate is thought to originate from the overlapping of the energy release from each of the 60 carbon atoms that result from the impact-induced fragmentation of the C_{60}^+ cluster ion. The second point is that the depth of energy deposition is approximately the depth of the sputter crater; this means that breaks in chemical bonds and modifications to the molecular structure of the sample occur in a region where the sample material will be sputtered away. The resultant surface after C_{60}^+ sputtering will present a composition that resembles the original composition, making it ideal for molecular depth profiling. The third point is that the C_{60}^+ primary ion fragments are not implanted deeply in the sample – as noted for the Ga^+ ion, which remains in the sample after the sputtering process (see Fig. 18.1).

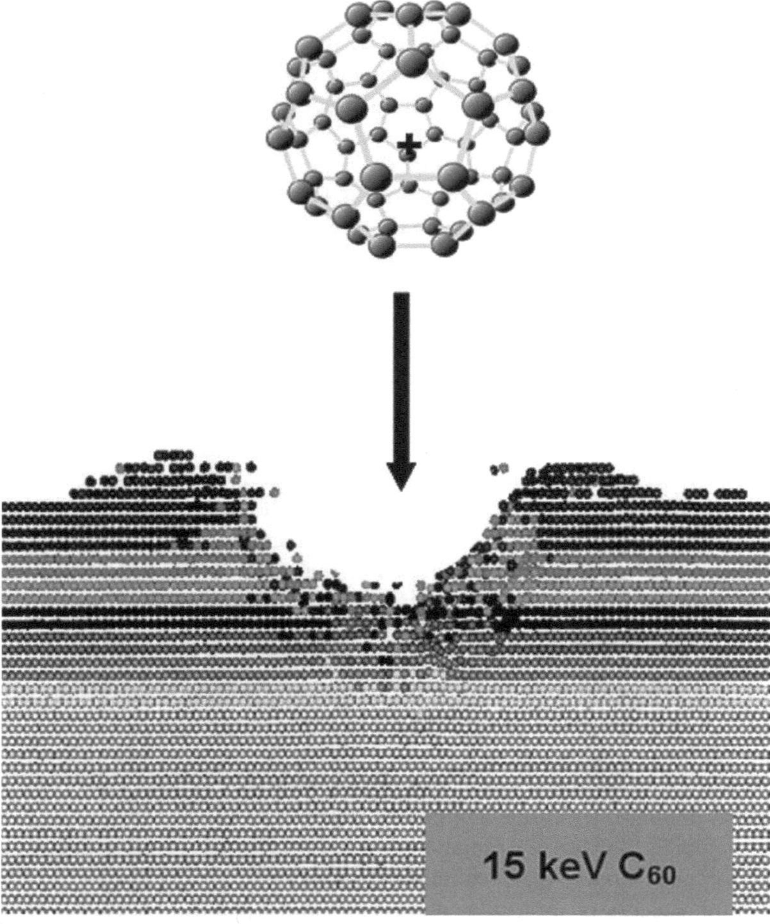

Fig. 18.7 Molecular dynamics simulation of impact of 15-kV C_{60}^+ ion into Ag after 29 ps

In fact, most of the C atoms from the C_{60}^+ primary ion are removed in the sputter process. However, recent C_{60}^+ depth profiles on model polymers now suggest that the eventual accumulation of carbon at the bottom of the sputter crater may be responsible for terminating successful TOF-SIMS molecular depth profiling [14]. The depth of this termination depends on the organic material used and the energy of the incident C_{60}^+ primary ions. However, molecular depth profiles from 1 to 3 μm can be practical for many samples. The fourth point is that the depth profile induces a slight roughening to the organic material consequent to the very high sputter rate for each impacting ion: this has been observed by comparisons of atomic force microscopy (AFM) measurements of the outer surface and the crater bottom of the C_{60}^+ depth profile of thick polymer films [14].

The data acquisition of 3D images uses a combination of analytical protocols used for 2D retrospective TOF-SIMS imaging discussed earlier and the interleaved TOF-SIMS depth profiling that is highly successful with inorganic multilayer samples. The microprobe primary ion gun is rastered over the sample and data for all secondary ions, produced at each pixel, are recorded in the matrix of a raw file, together with the coordinates of the x and y for pixel position, t for the flight time to the detector, and c for the sputter cycle. After recording the first 2D image, the user-defined length of rastered sputtering of the organic material by the C_{60} ion gun will remove the desired thickness of the sample. The process is then repeated for the next sputter cycle to produce a series of image slices as a function of depth into the material. The user then retrospectively displays the data, using the same software routines used in the 2D analysis, except that the axis of depth is added to the display.

For the visualization of 3D structures, the software function of iso-surface structure modeling can be highly useful. In this routine, the user selects a maximum and minimum intensity for a particular mass or fingerprint of masses. These intensity thresholds are then used to determine which pixels will be grouped together from the x, y, and c coordinates matrix of the 3D raw data file, to construct a 3D model. The iso-surface models can then be displayed using color-coding and a color opacity level. The user can then overlay as many models as desired in a 3D rendering of the sample. The overlaid models can be displayed on the computer screen, using 3D axis rotation to identify the 3D locations of the coincidence or anticoincidence of different molecular species. An example of this 3D analysis is presented later in this chapter, with reference to the analysis of drugs in a drug-eluting polymer coating used on a medical stent.

18.2.6 Sample Preparation

Because of the excellent surface sensitivity of the TOF-SIMS technique – i.e., generally one to two atomic layers or one molecular layer – the cleanliness of the sample preparation is of extreme importance. It is beyond the scope of this chapter to cover all the important details regarding sample cleanliness; instead, the reader is referred to an excellent treatise on this topic by Reich [15].

The preparation of samples for MS imaging of organic and biological samples deserves some additional comments. The first covers the imaging of tissue samples. Most of these samples are thin, cross-sectioned samples prepared with a cryo-microtome. The techniques are similar to the sample preparation techniques used for MALDI cross-section samples, and the reader is referred to earlier chapters in this book (Chaps. 1–8). However, it cannot be stressed enough that the cleanliness of the microtome blade will have a dramatic impact on the level of contamination detected in the TOF-SIMS spectra. The use of a freshly cleaned glass blade is therefore recommended.

Following cross-sectioning, the sample can be introduced into the TOF-SIMS vacuum system. All commercial TOF-SIMS instruments now have optional controlled temperature sample loading locks for sample evacuation and controlled temperature analysis stages. These sample handling systems allow the samples to be introduced into the ultrahigh vacuum chamber and analyzed without the loss of volatile sample components; it does so by using liquid nitrogen cooling hardware, together with heating, to actively control the sample temperature. The addition of heating and liquid nitrogen cooling allows the sample temperature to be maintained at an optimal temperature. It also prevents the buildup of ice on the sample surface, which can result from the condensation of water vapor from inside the vacuum system onto a cold sample surface.

The second issue is how the effects of different ion yields that complicate the intensity levels observed in TOF-SIMS imaging can be ameliorated. This is confounded by the desire to increase the secondary ion yield to achieve higher sensitivity and better spatial resolution based on better imaging statistics. One technique is based on the addition of typical MALDI organic matrix species, to enhance the SIMS analysis. The initial work of Wu and Odom was based on the mixing of peptides and oligonucleotides with 2,5-dihydroxybenzoic acid (DHB). The results indicated that ion yield enhancements for species up to m/z 10,000, ranging from 25- to 1,000-fold, could be observed [16]. The early work of Wu did not, however, address the improved imaging made available by this technique. Other research groups, including Heeren et al., used electrospray to apply extremely fine (<1 μm diameter) crystals of DHB on cryosections of the cerebral ganglia of the freshwater snail *Lymnaea stagnalis*. Good images were obtained using an $^{115}In^+$ LMIG primary ion source to obtain molecular maps of cholesterol and the neuropeptide APGWamide, with a spatial resolution to <3 μm and a high mass sensitivity to 2,500 Da [17]. The addition of DHB appears to enhance the ion formation of molecules and molecular fragments sputtered by the primary ion beam. It is noteworthy that the mass range of the detected molecular species is limited to much less than 10,000 Da, probably reflecting the limitation of the sputtering process in desorbing molecules of higher molecular weights.

A similar enhancement process is based on the coating of organic materials and tissue cross sections with ultrafine gold and silver particles. Bertrand et al. explored the use of a 2-nm-thick gold metallization to enhance the ionization probabilities for a series of thick polymers and additives with molecular weights in the 1,000- to 5,000-Da range. The results indicated dramatic increases in ion yield of more than two orders of magnitude and enhancements in the number of both gold cationized molecular ions and noncationized ions, using a 15-keV Ga^+ LMIG primary ion beam. The enhancement was dependent on the polymer species [18].

Ron Heeren et al. combined the deposition of gold with the application of traditional MALDI matrix crystals to enhance the imaging of tissues and cells. For the analysis of tissue cross sections prepared by cryomicrotomy, the deposition of gold resulted in gold islands on the order of 100 nm, as observed with AFM. For the analysis of single neuroblastoma cells grown on glass slides and then frozen and dried, DHB was applied by electrospray to produce small (0.3–1 μm) matrix crystals on the surface of the cells. The imaging of these cells, called matrix-enhanced (ME)-SIMS, produced enhanced lipid molecular ion images of the single cells.

The SIMS signal intensities for this sample preparation produced excellent intact molecular ion imaging of phosphatidylcholine and sphingomyelin at the cellular level. A second group of cells was coated by electrospray with DHB and then subsequently coated with 1-nm films of gold. The imaging of these cells, called metal-assisted (MetA) SIMS, showed strongly enhanced signal intensities for higher spatial resolution imaging. In addition, additional peaks in the mass range of m/z 600–1,300 were observed. The spectral resolving power of the MetA-SIMS was about 1 μm, as a result of the convolution of the spatial resolution of the TOF-SIMS instrument and the metal and matrix-assisted signal intensity needed for imaging. The deposition of the thin layer of gold appeared also to decrease the surface diffusion and improve the spatial resolution, compared to the ME-SIMS sample preparation [19].

18.3 Applications of TOF-SIMS Imaging

18.3.1 Tissue Cross-Section Analyses

The wide-ranging success of TOF-SIMS for imaging inorganic and semiconductor materials has led several research groups to explore the potential of TOF-SIMS in imaging cross sections of tissue samples and, ultimately, to focus on subcellular features in cells. The aim of this chapter is not to review all these efforts but rather to provide a selection of examples that illustrate the success and future potential stemming from these research efforts. The preparation of thin cross-section tissue materials has been developed for scanning electron microscopy, transmission electron microscopy, and fluorescence microscopy techniques, and the reader is referred to one of several texts for details of this sample preparation methodology [20].

One of the early efforts to examine TOF-SIMS images of tissue samples was by Wu et al. at Lawrence Livermore National Laboratory, University of California. The sample used was a microtome cross section of a 16-day-old mouse embryo that had been embedded in paraffin. Total ion TOF-SIMS images and selected ion images of the selected areas of the embryo are shown in Fig. 18.8. The TOF-SIMS total ion images of the different organs within the cross section, as well as the selected ion images of different structures in the embryo, suggested that TOF-SIMS could be a useful MS tool in characterizing tissues.

A primary goal of Wu's early research was to explore if TOF-SIMS could differentiate between organs, based on the mass spectra of different regions. The mass spectra in this research were not acquired with a higher mass primary ion cluster source, which would have resulted in abundant low mass fragment peaks, but a lack of the desired molecular peaks. The resultant spectra therefore showed an overlay of the TOF-SIMS fragmentation patterns from many of the different molecular species in each region, rendering an overall "similar" look across the different spectra. Wu et al. [21] used the approach of principal component analysis (PCA) to look for components in the spectra that could separate the areas based on interpretations of multivariate analysis. In Fig. 18.9, the intensity of the Ca^+ or Na^+ images from bone

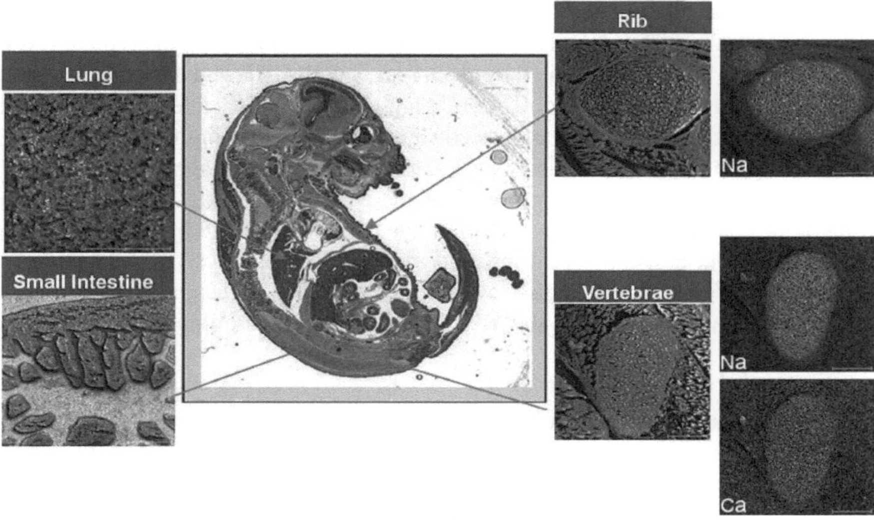

Fig. 18.8 Optical image of the cross section of a 16-day-old mouse embryo embedded in paraffin, with selected area total ion and selected ion TOF-SIMS images

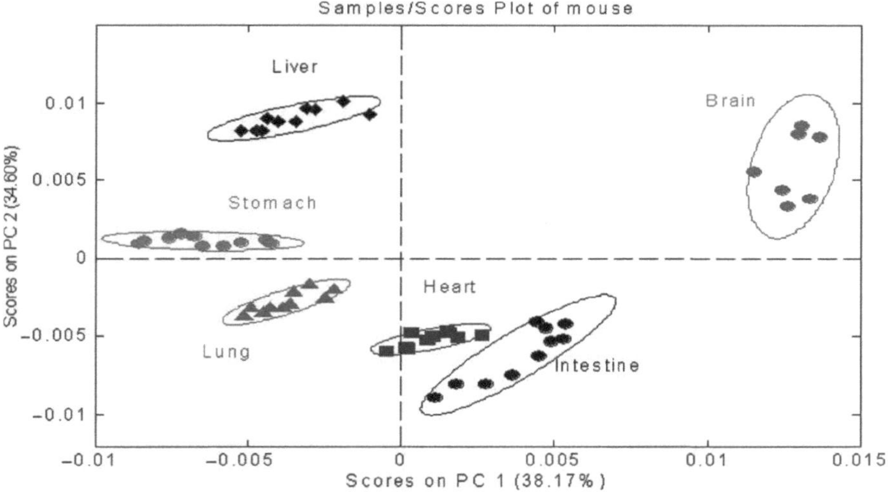

Fig. 18.9 Principal component analysis (PCA) separation of mass spectra from regions of six different organs in a 16-day-old mouse embryo. *PC* principal component

structures can be directly correlated with the optical image. For soft tissues, no single peak can uniquely define an organ structure. In PCA, a fingerprint of different peaks is defined as 1 eigenvector for the matrix of mass and intensity data. By plotting the loading scores for the first PC – i.e., the component that includes the highest variance in the matrix of data – versus the loading scores for the second PC as a function of the x and y locations in the image, a multivariate analysis can uniquely separate the regions of the liver, stomach, lung, heart, intestine, and brain in the embryo.

As encouraging as these data are regarding the future of TOF-SIMS imaging of tissue samples, several questions remain. The first question is the identity of the molecular constituents in each area. These data show only the pattern of low mass fragments, not molecular species. The second question is the origin in the cell of these fragment species. The cell membranes are primarily lipid structures; do the TOF-SIMS data show mass fragment differences in the lipid structures or differences in molecules attached to the lipid structures, or do the TOF-SIMS data also sample some of the intracellular material that allows differentiation of the organs [19]?

To further the molecular specificity and spatial resolution of TOF-SIMS imaging of tissue cross-section samples, Laprévote et al. [22] analyzed a 15-µm-thick cross section of a rat brain cut at a temperature of ~–20°C, stored on a stainless steel mount at ~–80°C, and vacuum-dried at room temperature immediately before TOF-SIMS analysis. Based on the expected increase in the secondary ion yield resulting from the use of an Au_3^+ primary cluster ion source, the sample was not coated with a matrix such as DHB or a nanometer film of gold. Fig. 18.10 shows an excellent complementary image of phospholipid ions and cholesterol ions with a 30-min stage rastered image over an 8 mm×8 mm area, with a pixel resolution of 62.5 µm×62.5 µm. The different regions of the rat brain can be clearly differentiated in this image.

To illustrate the improved imaging with the Au_3^+ primary cluster ion source, a region of the corpus callosum was imaged (Fig. 18.11) with a 500 µm×500 µm field of view and with 1.95 µm×1.95 µm pixel dimensions, by raster-scanning the primary ion source. The enhanced ion yield of the Au_3^+ allows excellent imaging with similar spatial resolution to the MetA-SIMS data of Heeren et al. It should be noted that the MetA-SIMS technique has produced images for ions in the 1,000- to 1,500-Da

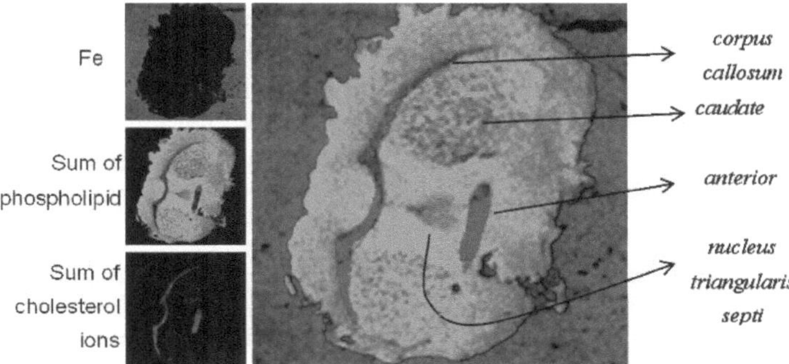

Fig. 18.10 Selected ion images obtained for stainless steel substrate, sum of characteristic phospholipid ions, sum of cholesterol ions, and a color overlay of the three images

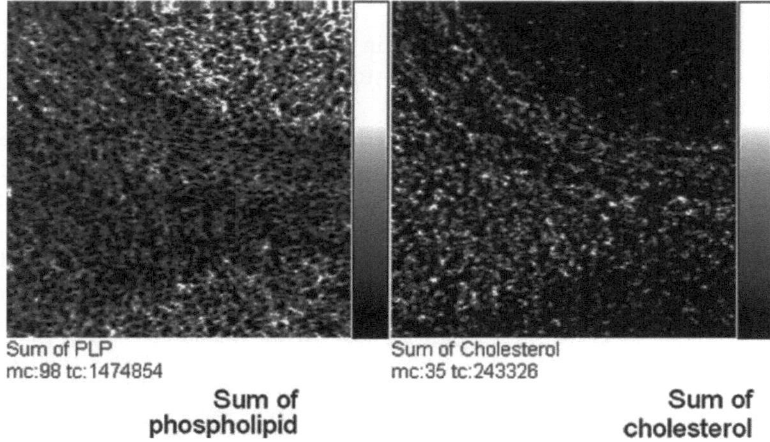

Sum of phospholipid   **Sum of cholesterol**

Fig. 18.11 Selected ion images obtained for the sum of characteristic phospholipid (*PLP*) ions and the sum of cholesterol ions from a region of the corpus callosum in the cross section of a rat brain. The data were acquired with an Au_3^+ primary cluster ion source raster-scanned over a $500\,\mu m \times 500\,\mu m$ field of view and with $1.95\,\mu m \times 1.95\,\mu m$ pixel dimensions

mass range, but the useful mass range for imaging with the Au_3^+ primary cluster ion source for tissue cross sections has been reported to be about 900 Da [22].

18.3.2 Imaging of Disease Biomarkers

The previous results demonstrate TOF-SIMS imaging spatial resolution capabilities for tissue cross sections approaching 1 μm, but the data have a limitation regarding useful mass range for imaging of <2,000 Da. These results have prompted a more narrow focus for many TOF-SIMS research groups. For many biological and medical applications based on MS, the identification of proteins and other large biomolecules is being pursued using MALDI and LC-MS/MS in conjunction with electrospray ionization techniques. The localization of metabolites and therapeutic drugs, however, represents an analytical opportunity that may more closely match the strengths of TOF-SIMS. The imaging of metabolites that are related to disease states of cells or an entire living animal has recently shown great promise; one recent application published by Laprévote et al. demonstrates the value of TOF-SIMS for MS imaging of the biomarkers of Fabry's disease [23]. Fabry's kidney disease is a lysosomal genetic disease caused by a dysfunction of α-D-galactosidase-A (α-GALA). It is linked to the X chromosome and affects 1 in 120,000 humans. The disease promotes the accumulation of glycosphingolipids Gb_3 (trihexosylceramide) and Ga_2 (digalactosylceramide) in kidney cells. The mass spectra of Gb_3 and Ga_2 are shown in Fig. 18.12, and images of Gb_3 and Ga_2 from a biopsy tissue sample of a human kidney are shown in Fig. 18.13. The imaging data were acquired

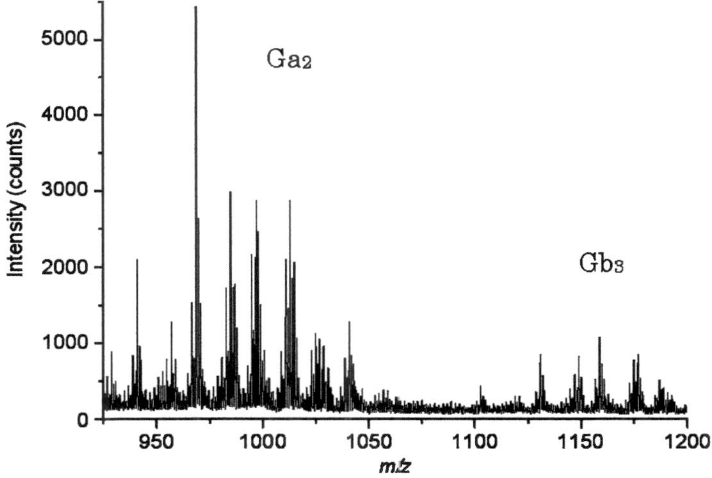

Fig. 18.12 Mass spectral data obtained with Bi_3^+ primary ion source from muscle cells for Ga_2 and Gb_3

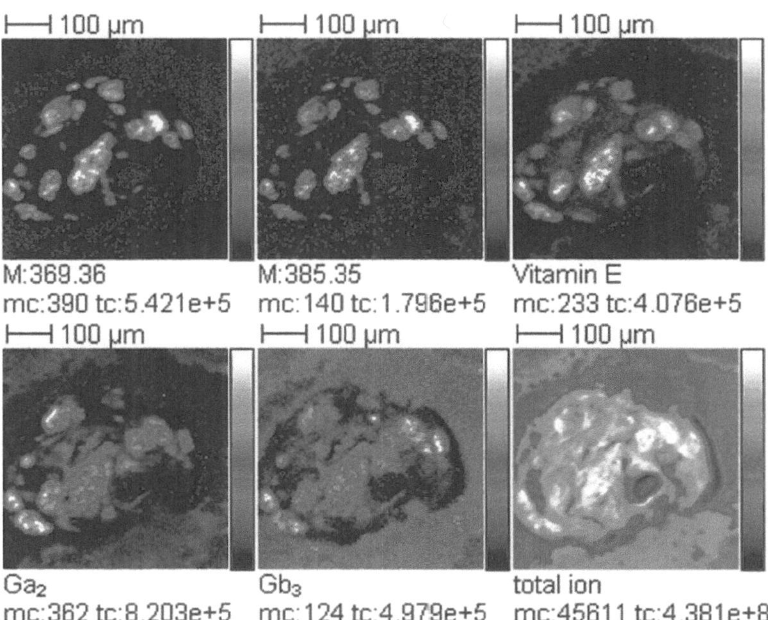

Fig. 18.13 Images of cholesterol at m/z 369 $(M+H-H_2O)^+$, m/z 385 $(M-H)^-$, vitamin E, Ga_2, and Gb_3, and total ion image obtained with Bi_3^+ primary ion source from muscle cells of Duchenne muscular dystrophy-afflicted subjects

in 30 min with a raster-scanned Bi_3^+ primary ion gun over a $500\,\mu m \times 500\,\mu m$ field of view and with $1\,\mu m \times 1\,\mu m$ pixel dimensions. The images in Fig. 18.13 show a colocalization of Gb_3 and Ga_2, as well as a colocalization with vitamin E, cholesterol, and cholesterol sulfate (data not shown). The authors claim an image spatial resolution for Gb_3 and Ga_2 of $1\,\mu m$. Similar imaging results of biomarkers for muscle cell samples from mice and humans suffering from Duchenne muscular dystrophy have shown the localization of cholesterol, vitamin E, and fatty acids of different chain lengths in regions of different cell types [24]. These valuable results point to the unique analytical capability of TOF-SIMS for cellular and subcellular MS imaging in supporting medical research.

18.3.3 Imaging of Drugs with Controlled Drug-Release Coatings

Presently, most therapeutic drugs have a mass <1,000 Da. Historically, most drug treatments have depended on oral ingestion or intravenous administration. The recent increase in the use of implanted medical devices has added a new dimension to the requirements for drug delivery. Despite the best efforts of material scientists to develop materials and surface coatings that are "biocompatible," the human body has a natural tendency to attack any implanted materials as foreign bodies; this is particularly evident with the increased use of stents to treat cardiovascular disease. Although the expandable stent has become a common tool in treating blockages in the heart, complications arising from the body enveloping the stent and creating new blockages in the artery causes anxiety in patients and consternation in the medical community. To prevent this process of restenosis, stent manufacturers are now developing drug-release coatings that can be applied directly to the stent. A key factor in the development of these coatings is an understanding of the 3D concentration of the drug and the pathways for drug elution from the coating. 3D MS imaging of the drug in the coating, as well as the periodic analysis of this drug composition as a function of elution of the drug, is an exciting new application of TOF-SIMS.

In a recent study by Fisher et al. [25] at Physical Electronics in Chanhassen, MN, the 3D compositional analysis of the drug Rapamycin in a soluble coating of poly(lactic-co-glycolic acid) deposited on a test coupon was obtained with TOF-SIMS. 2D images of the drug and coating were obtained at a series of depths from the outer surface to a depth of $3\,\mu m$ into the coating, using 20-keV C_{60}^+ sputter depth profiling. The resultant collection of 2D images was combined using iso-surface 3D image display software. The results are displayed in Fig. 18.14.

To more fully understand the structure shown in Fig. 18.14, the opacity of the green iso-surface model, poly(lactic-co-glycolic acid), was reduced to allow the viewer to see into the coating in Fig. 18.15 [25].

This study illustrates the potential for 3D TOF-SIMS imaging of organic, biomaterial, and tissue samples, using the C_{60}^+ sputter depth profiling technique. Further studies are underway to explore the benefits of this approach with other materials, especially tissue samples.

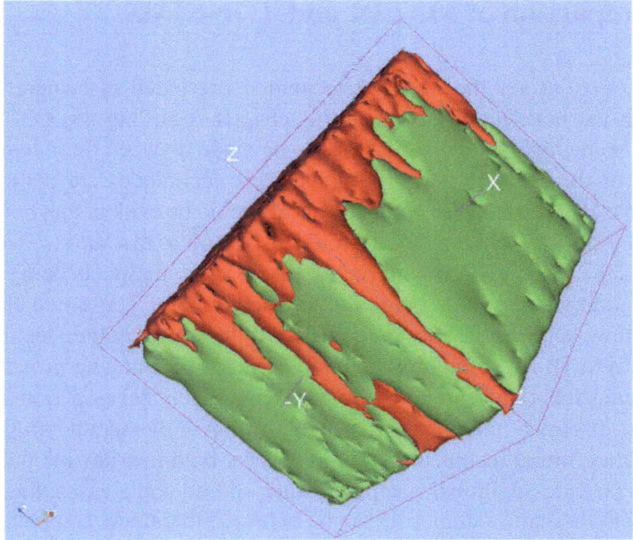

Fig. 18.14 The 3-D intensity of the drug Rapamycin (*red*) and the soluble coating poly(lactic-*co*-glycolic acid) (*green*), for a 200 μm width × 200 μm length × 2.8 μm depth

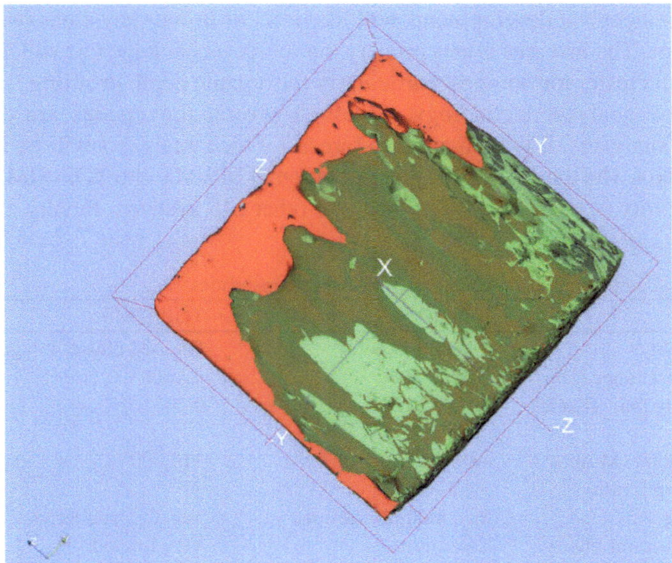

Fig. 18.15 The 3D intensity of the drug Rapamycin (*red*) and the soluble coating poly(lactic-*co*-glycolic acid) (*green*) for a 200 μm width × 200 μm × 2.8 μm depth. The opacity of the coating model (in *green*) has been reduced to allow the multiple channels in the coating to be more visible

18.4   Comparison of MALDI and TOF-SIMS

The several sections on the MALDI technique presented elsewhere in this text, together with the material discussed in this chapter regarding the TOF-SIMS technique, allow some high-level comparisons to be made between these two techniques. A summary of the characteristics by which these techniques are compared is presented in Table 18.2. From these comparisons, it should be evident that both techniques are powerful tools for MS imaging; it should also be clear that the two techniques are complementary in many ways, with the two most notable aspects being the imaging spatial resolution and the mass range of the detected ions. These two characteristics will likely allow application developments from these two techniques that are complementary. MALDI has a required position as a tool for imaging proteins and other macromolecules found in cells, and TOF-SIMS will likely develop as an imaging tool that examines disease biomarkers and the localization of therapeutic drugs. It can also be expected that further technical developments for both instruments will allow each technique to provide additional complementary and, in some cases, directly comparable data from the same samples. It can be expected that many research laboratories will use both these techniques in MS imaging.

18.5   Future Developments for TOF-SIMS

Significant technical developments will likely occur in four key areas over the next several years. The first area is new primary ion source technology; this will likely focus on different cluster ion sources, particularly for organic depth profiling. The second area is new analyzer technology. Although several universities are developing prototype analyzers, it is premature to forecast which analyzers will be introduced commercially. The third area is the incorporation of MS/MS into TOF-SIMS systems. This capability is widespread for almost all other MS systems. Finally, there is the

Table 18.2 Comparison of MALDI and TOF-SIMS capabilities

Capability	MALDI	TOF-SIMS
Incident probe	Focused laser	Focused cluster ion gun
Practical mass range (Da)	>100,000	<2,000
Spatial resolution, organics (μm)	30–50	0–5
Mass resolution (M/Δm)	~10,000	~15,000
Sampling depth (μm)	1	<1
3-D imaging	Yes, multiple sections	Yes, C_{60}^+ sputtering
Molecular spectral library	Extensive	Very limited
Usual imaging approach	Stage mapping	Scanned ion gun and stage mapping

Note: These are not theoretical limits, but widely used capabilities. Some laboratories have published findings from equipment showing enhanced capabilities with both techniques

incorporation of software to quickly and conveniently interpret and output enormous amounts of data obtained in 2D and 3D raw files. It is not unreasonable to expect that both MALDI and TOF-SIMS will continue to grow in terms of performance, commercial offerings, and laboratory use around the world.

References

1. Vickerman JC, Brown A, Reed NM (1989) Secondary ion mass spectrometry. Oxford University Press, New York
2. Vickerman JC, Briggs D, eds. (2001) ToF-SIMS: surface analysis by mass spectrometry. IM Publications, Chichester
3. Postawa Z, Czerwinski B, Szewczyk M, Smiley EJ, Winograd N, Garrison BJ (2004) J Phys Chem B 108:7832–7838
4. Delcorte A, Garrison BJ (2004) J Phys Chem B 104:6785–6800
5. Cameca nanoSIMS 50. Cameca, Gennevilliers, France
6. Physical Electronics, Chanhassen, MN, USA: unpublished results
7. ION-TOF. GmbH, Münster, Germany
8. Physical Electronics USA, Chanhassen, MN, USA and ULVAC-PHI Inc., Chigasaki, Japan
9. SurfaceSpectra, Manchester, UK
10. Gillen G, Roberson S (1998) Rapid Commun Mass Spectrom 12:1202
11. van Stipdonk MJ (2001) In: Vickerman JC, Briggs D, eds. ToF-SIMS: surface analysis by mass spectrometry. IM Publications, Chichester, pp 309–345
12. Ionoptika Ltd., Epsilon House, Chilworth Science Park, Southampton S016 7NS, UK
13. Sanada N. ULVAC-PHI Inc. Chigasaki, Japan: unpublished results
14. Fisher G. Physical Electronics Inc, Chanhassen, MN, USA: unpublished results
15. Reich DF (2001) In: Vickerman JC, Briggs D, eds. ToF-SIMS: surface analysis by mass spectrometry. IM Publications, Chichester, pp 113–135
16. Wu KJ, Odom RW (1996) Anal Chem 68:873–832
17. Altelaar AFM, Van Minnen J, Jimenez CR, Heeren RMA, Piersma SR (2005) Anal Chem 77:735–741
18. Delcorte A, Bour J, Aubriet F, Muller J-F, Bertrand P (2003) Anal Chem 75:6875–6885
19. Altelaar AFM, Klinkert I, Jalink K, de Lange RJP, Adan RAH, Heeren RMA, Piersma SR (2006) Anal Chem 78:734–742
20. Severs N, Shotton D, eds. (1995) Rapid freezing, freeze-fracture, and deep etching. Wiley-Liss, New York
21. Private communication with Kwang Jen J Wu, Lawrence Livermore National Laboratory, University of California
22. Touboul D, Halgand F, Brunelle A, Kersting R, Tallarek E, Hagenhoff B, Laprévote O (2004) Anal Chem 76:1550–1559
23. Touboul D, Roy S, Germain DP, Chaminade P, Brunelle A, Laprévote O (2007) Int J Mass Spectrom 260:158–165
24. Touboul D, Brunelle A, Halgand F, De Le Porte S, Laprévote O (2005) J Lipid Res 46:1388
25. Fisher GF. Physical Electronics Inc., Chanhassen, MN, USA

Index

MIX
Papier aus verantwortungsvollen Quellen
Paper from responsible sources
FSC® C105338

If you have any concerns about our products,
you can contact us on
ProductSafety@springernature.com

In case Publisher is established outside the EU,
the EU authorized representative is:
Springer Nature Customer Service Center GmbH
Europaplatz 3, 69115 Heidelberg, Germany

Printed by Libri Plureos GmbH
in Hamburg, Germany